Enjoying Organic Chemistry, 1927–1987

PROFILES, PATHWAYS, AND DREAMS

Autobiographies of Eminent Chemists

Jeffrey I. Seeman, Series Editor

Enjoying Organic Chemistry, 1927–1987

Egbert Havinga

American Chemical Society, Washington, DC 1991

Library of Congress Cataloging-in-Publication Data

Havinga, E. (Egbert). 1909–1988.
  Enjoying Organic Chemistry, 1927–1987 / E. Havinga.

    p.   cm.(Profiles, pathways, and dreams)

  Includes bibliographical references and index.

  ISBN 0–8412–1774–2.—ISBN 0–8412–1800–5 (pbk.)

    1. Havinga, E. (Egbert), 1909–1988. 2. Chemists—
Netherlands—Biography.   3. Chemistry, Organic—
History—20th century.
   I. Title. II. Series.

QD22.H317A3 1990
540'.92—dc20
[B]   90–45265
                                        CIP

The paper used in this publication meets the minimum requirements of American National Standard for Information Sciences—Permanence of Paper for Printed Library Materials, ANSI Z39.48–1984.

∞

PRINTED IN THE UNITED STATES OF AMERICA

Second printing 1992

Foreword

In 1986, the ACS Books Department accepted for publication a collection of autobiographies of organic chemists, to be published in a single volume. However, the authors were much more prolific than the project's editor, Jeffrey I. Seeman, had anticipated, and under his guidance and encouragement, the project took on a life of its own. The original volume evolved into 22 volumes, and the first volume of *Profiles, Pathways, and Dreams: Autobiographies of Eminent Chemists* was published in 1990. Unlike the original volume, the series was structured to include chemical scientists in all specialties, not just organic chemistry. Our hope is that those who know the authors will be confirmed in their admiration for them, and that those who do not know them will find these eminent scientists a source of inspiration and encouragement, not only in any scientific endeavors, but also in life.

M. Joan Comstock
Head, Books Department
American Chemical Society

Contributors

We thank the following corporations and Herchel Smith for their generous financial support of the series Profiles, Pathways, and Dreams.

Akzo nv

Bachem Inc.

E. I. du Pont de Nemours and Company

Duphar B.V.

Eisai Co., Ltd.

Fujisawa Pharmaceutical Co., Ltd.

Hoechst Celanese Corporation

Imperial Chemical Industries PLC

Kao Corporation

Mitsui Petrochemical Industries, Ltd.

The NutraSweet Company

Organon International B.V.

Pergamon Press PLC

Pfizer Inc.

Philip Morris

Quest International

Sandoz Pharmaceuticals Corporation

Sankyo Company, Ltd.

Schering–Plough Corporation

Shionogi Research Laboratories, Shionogi & Co., Ltd.

Herchel Smith

Suntory Institute for Bioorganic Research

Takasago International Corporation

Takeda Chemical Industries, Ltd.

Unilever Research U.S., Inc.

Profiles, Pathways, and Dreams

Titles in This Series

About the Editor

JEFFREY I. SEEMAN received his B.S. with high honors in 1967 from the Stevens Institute of Technology in Hoboken, New Jersey, and his Ph.D. in organic chemistry in 1971 from the University of California, Berkeley. Following a two-year staff fellowship at the Laboratory of Chemical Physics of the National Institutes of Health in Bethesda, Maryland, he joined the Philip Morris Research Center in Richmond, Virginia, where he is currently a section leader. In 1983–1984, he enjoyed a sabbatical year at the Dyson Perrins Laboratory in Oxford, England, and claims to have visited more than 90% of the castles in England, Wales, and Scotland.

Seeman's 80 published papers include research in the areas of photochemistry, nicotine and tobacco alkaloid chemistry and synthesis, conformational analysis, pyrolysis chemistry, organotransition metal chemistry, the use of cyclodextrins for chiral recognition, and structure–activity relationships in olfaction. He was a plenary lecturer at the Eighth IUPAC Conference on Physical Organic Chemistry held in Tokyo in 1986 and has been an invited lecturer at numerous scientific meetings and universities. Currently, Seeman serves on the Petroleum Research Fund Advisory Board. He continues to count Nero Wolfe and Archie Goodwin among his best friends.

Contents

Photographs

Preface

"HOW DID YOU GET THE IDEA—and the good fortune—to convince 22 world-famous chemists to write their autobiographies?" This question has been asked of me, in these or similar words, frequently over the past several years. I hope to explain in this preface how the project came about, how the contributors were chosen, what the editorial ground rules were, what was the editorial context in which these scientists wrote their stories, and the answers to related issues. Furthermore, several authors specifically requested that the project's boundary conditions be known.

As I was preparing an article[1] for *Chemical Reviews* on the Curtin–Hammett principle, I became interested in the people who did the work and the human side of the scientific developments. I am a chemist, and I also have a deep appreciation of history, especially in the sense of individual accomplishments. Readers' responses to the historical section of that review encouraged me to take an active interest in the history of chemistry. The concept for Profiles, Pathways, and Dreams resulted from that interest.

My goal for Profiles was to document the development of modern organic chemistry by having individual chemists discuss their roles in this development. Authors were not chosen to represent my choice of the world's "best" organic chemists, as one might choose the "baseball all-star team of the century". Such an attempt would be foolish: Even the selection committees for the Nobel prizes do not make their decisions on such a premise.

The selection criteria were numerous. Each individual had to have made seminal contributions to organic chemistry over a multidecade career. (The average age of the authors is over 70!) Profiles would represent scientists born and professionally productive in different countries. (Chemistry in 13 countries is detailed.) Taken together, these individuals were to have conducted research in nearly all subspecialties of organic chemistry. Invitations to contribute were based on solicited advice and on recommendations of chemists from five continents, including nearly all of the contributors. The final assemblage was selected entirely and exclusively by me. Not all who were invited chose to participate, and not all who should have been invited could be asked.

A very detailed four-page document was sent to the contributors, in which they were informed that the objectives of the series were

1. to delineate the overall scientific development of organic chemistry during the past 30–40 years, a period during which this field has dramatically changed and matured;

2. to describe the development of specific areas of organic chemistry; to highlight the crucial discoveries and to examine the impact they have had on the continuing development in the field;

3. to focus attention on the research of some of the seminal contributors to organic chemistry; to indicate how their research programs progressed over a 20–40-year period; and

4. to provide a documented source for individuals interested in the hows and whys of the development of modern organic chemistry.

One noted scientist explained his refusal to contribute a volume by saying, in part, that "it is extraordinarily difficult to write in good taste about oneself. Only if one can manage a humorous and light touch does it come off well. Naturally, I would like to place my work in what I consider its true scientific perspective, but . . ."

Each autobiography reflects the author's science, his lifestyle, and the style of his research. Naturally, the volumes are not uniform, although each author attempted to follow the guidelines. "To write in good taste" was not an objective of the series. On the contrary, the authors were specifically requested not to write a review article of their field, but to detail their own research accomplishments. To the extent that this instruction was followed and the result is not "in good taste", then these are criticisms that I, as editor, must bear, not the writer.

As in any project, I have a few regrets. It is truly sad that Egbert Havinga, who wrote this volume, and David Ginsburg, who translated another, died during the development of this project. There have been many rewards, some of which are documented in my personal account of this project, entitled "Extracting the Essence: Adventures of an Editor" published in CHEMTECH.[2]

Acknowledgments

I join the entire chemical community in offering each author unbounded thanks. I thank their families and their secretaries for their contributions. Furthermore, I thank numerous chemists for reading and reviewing each volume, for lending photographs, for sharing information, and for providing each of the authors and me the encouragement to proceed in a project that was far more costly in time and energy than any of us had anticipated.

I thank my employer, Philip Morris USA, and J. Charles, R. N. Ferguson, K. Houghton, and W. F. Kuhn, for without their support Profiles, Pathways, and Dreams could not have been. I thank ACS Books, and in particular, Robin Giroux (acquisitions editor), Karen Schools Colson and Kathleen Strum (production managers), Janet Dodd (senior editor), Joan Comstock (department head), and their staff for their hard work, dedication, and support. Each reader no doubt joins me in thanking 24 corporations and Herchel Smith for financial support for the project.

I thank my children, Jonathan and Brooke, for their patience and understanding; remarkably, I have been working on Profiles for more than half of their lives—probably the only half that they can remember! I want to thank also a group of friends who were especially supportive during the production of this volume. Finally, I again thank all those mentioned and especially my family, friends, colleagues, and the 22 authors for allowing me to share this experience with them.

JEFFREY I. SEEMAN
Philip Morris Research Center
Richmond, VA 23234

November 11, 1990

[1] Seeman, J. I. *Chem. Rev.* **1983**, *83*, 83–134.
[2] Seeman, J. I. *CHEMTECH* **1990**, *20*(2), 86–90.

Editor's Note

I always very much enjoyed conversing with Egbert Havinga. He had a way of making me feel good; even his goodbyes had a special ring of sincerity and honesty. The optimism Havinga had radiated in the last few months of his life had shielded me and others from the reality of his illness. On the morning of November 2, 1988, H. J. C. (Harry) Jacobs phoned and told me that Havinga's health was failing rapidly, but that he was still very interested in our project. He wanted to know the status of the Profiles series but did not want to bother me with asking. Havinga, Jacobs explained, "wanted to die in a friendly way, not disturbing people". My response, sent by telefax within the hour, was one of the most difficult I had ever written. Twenty days later, he was gone.

Havinga was a man who directed attention away from himself. His most self-effacing act was to keep the details of his fatal illness from his friends until the final moments. When I read the initial draft of this volume, I found much evidence of his selflessness in statements such as "Dallinga in 1951", "Kwestroo who synthesized", and "J. A. van der Linden and E. C. Wessels studied", but the accompanying references revealed that these were all Havinga students. It was Havinga's way of shining the light onto others, in his special gentlemanly manner.

Human beings were very important to Havinga, and in addition to being interested in their science, he was interested in them. According to a colleague, "Frequently, people would seek his advice on many matters, scientific, personal, and administrative. [Havinga] was very careful in formulating his responses, not wanting to impose his opinion, not wanting to make too much of a fuss ... ", but wanting to offer as much as he could to be helpful. These, among his many fine traits, endeared Havinga to his friends and colleagues.

Since his earliest days as a professor of organic chemistry at Leiden University, Havinga paid much attention to teaching. He considered teaching of extreme importance, especially teaching beginning students. His students attest to the quality of Havinga's lectures, which were especially popular and attracted numerous auditors.

Havinga's science ranged from physical organic to pyrogenic chemistry. He worked in many diverse areas, and also published in many different specialized journals, in addition to publishing frequently in *Recueil des Travaux Chimiques des Pays–Bas* (the *Journal of the*

Royal Netherlands Chemical Society). Many readers of Havinga's autobiography will be pleasantly surprised at his (Havinga would probably say his students') broad contributions.

Together with his esteemed colleague Luitzen Oosterhoff, Havinga recognized and subsequently published in 1961 "the first hint" of orbital symmetry control of concerted organic reactions. For me, personally, seeing the figure that appears on page 36 of this book brought back vivid and telling memories. As a young graduate student in William G. Dauben's laboratory in 1967, I had eagerly devoured a tiny paperback booklet, the Ph.D. thesis of Oosterhoff's student Van der Lugt, to understand what was influencing the photochemical reactions of the 1,3-cyclohexadienes I was studying. That figure and many of Havinga's papers served as the key references for my research.

As an academic chemist, Havinga enjoyed his frequent scientific contacts with industrial laboratories. And clearly, the Dutch chemical industry was proud of its association with Havinga; at least three corporations that supported the Profiles series had connections with him. Havinga was also happy about two additional marks he left at Leiden: his role in establishing chairs in theoretical organic chemistry and biochemistry in the early 1950s, the latter being the first of its kind in The Netherlands. In addition, Havinga spoke with delight on the planning of the new chemistry laboratories. I still recall the majesty of the Gorlaeus Laboratories when I first saw the beautiful complex. Havinga enthusiastically took part in its design, "stimulated by his thorough interest and feeling for art and architecture," according to a friend.

Havinga had many interests in addition to science. His attention included music—he played the piano with expertise—photography, sports, and nature. He strongly admired nature's beauty. Earlier than many others, he recognized the detrimental effect of human activities on the environment. And he had a remarkable reputation as a strong, very strong indeed, tennis player, as many participants of the Bürgenstock and other conferences will attest.

It brought Havinga considerable comfort that Harry Jacobs would complete the arduous and significant final tasks associated with this volume. This comfort was well placed, for Jacobs has responded immediately to all of my many requests.

Over the past 2 years, I have remained in infrequent but regular contact with Mrs. Havinga, a most lovely lady. Her feelings of continued sadness regarding her husband's passing are shared by many, for Egbert Havinga was a true gentleman, a kind individual, a good friend, and a seminal scientist.

November 11, 1990

Enjoying Organic Chemistry, 1927–1987

Egbert Havinga

The Early Years: Doctoral Research and Study

Many organic chemists, with their characteristic love of doing experimental investigations with their own hands, may feel—as I do—that the most enjoyable period of their life with chemistry has been that of the research for obtaining the doctor's degree. There was almost undisturbed concentration on the experiments: planning, doing, and interpreting. One profited from daily discussions with colleagues in the same phase of development. There existed unique social freedom, little hindrance by representational duties, administration, correspondence, etc. This happy period of concentration on my own research—which, of course, had its innate difficulties—occurred in the years before World War II at the University of Utrecht, where I started chemistry studies in 1927.

The choice in favor of chemistry—over the other favorite, Latin—had been made after finishing high school (*Gymnasium* at Amersfoort). My decision was not prompted by the traditional adventures with explosives in the cellar at home, but rather by the romantic expectation that the study of chemistry might lead to some understanding of the how's and why's in this world. Because the roots of preferences in the subject of later research developed largely in the period before 1940, it seems appropriate to describe the atmosphere at the university and the faculty of science half a century ago.

Students at that time had practically unlimited freedom to work *or* not to work and to spend their time with different activities. In accord with the European tradition, there were no more than three professors of chemistry at the University of Utrecht. They were excellent scientists and esteemed personalities: Ernst Cohen, outstanding pupil of Van't Hoff; Hugo R. Kruyt, one of the founders of modern colloid chemistry; Leopold Ruzicka; and his successor, Fritz Kögl, both famous organic chemists. The last two scientists had been invited to come from Switzerland and Germany, respectively, in order to stimulate the study of organic chemistry of natural products, which was then underdeveloped in the Netherlands.

Organic chemistry was in a period of heroic development: vitamins and hormones were being isolated, their structures established by sophisticated chemical degradation, and their syntheses accomplished in rapid succession. No wonder that I chose to specialize with Kögl, who was engaged in fascinating research on the then recently discovered plant growth hormones. He was assisted by very able co-workers like A.J. Haagen Smit (later a pioneer of smog research at California Institute of Technology) and H. Veldstra (later to become my colleague in biochemistry at Leiden University).

3

The author as a young boy dressed in what was at that time (around 1917) the highly popular sailor suit.

The second specialization, theoretical physics, at that time formed a somewhat unorthodox complement to organic chemistry. It was motivated by the inspiring lectures of H. A. Kramers, who made the fundamentals of quantum mechanics exciting even to an experimental organic chemist and prompted me to include a year of physics in my curriculum.

Experimental training became interesting in the last stages of the doctoral study, in which one was given special tasks mostly related to the "real research" directed by the professor. In my case, this was the synthesis of compounds required for the investigations of plant growth hormones. One of these syntheses involved a provoking experience. Crystallization from a supersaturated solution occurred only after I had set the solution aside in a closed flask for more than half a year. The eventual result was a successful preparation of the compound (one of the dihydroxyglutaric acid isomers). More importantly, it gave me a lasting interest in the problem of how crystallization can start at all with relatively large molecules of low symmetry in the absence of effective nuclei.

After the final (doctoral) examination, real life with chemistry began: thesis research. I had obtained an assistantship with Kögl, an eminent organic chemist with biological interests, who came from the

F. Kögl (1897–1959), professor of organic chemistry, University of Utrecht, 1930–1959.

renowned schools of Windaus and Hans Fischer. Later in his life he was to experience great disappointments because of what must have been falsifications by his long-term co-worker H. Erxleben, whom he valued highly for her experimental virtuosity. But in the 1930s he still was in the rapidly rising part of his scientific life-curve, full of energy and enthusiasm, working at the front line of bioorganic chemistry. Professor and Mrs. Kögl gained general respect and sympathy on account of their rejection of the ideas and behavior of the Nazis. This nonscientific aspect deserves explicit mention because the difficult situation during the occupation of the Netherlands for a man like Kögl, German by birth and education, cannot easily be evaluated.

Returning to the prewar chemical story: Kögl had developed the idea that "asymmetric synthesis"—since the days of Pasteur often considered the prerogative of living systems—is related to the fact that in living organisms reactions generally occur in interface layers (membranes or surfaces of enzyme complexes). He proposed that I would continue this research along the line initiated by A. A. Höchtlen, who then worked as a postdoctoral fellow at Utrecht. Of course Höchtlen had experienced quite a few difficulties. He started to do reactions with monomolecular layers that were spread on water and subsequently transferred onto glass plates by the then-novel Langmuir–Blodgett technique.

Even today the proposed thesis research could be judged a risky enterprise for both student and mentor. This was far more the case half

Professor Kögl (center) with his group at Utrecht University, around 1936. The author is to his left.

a century ago, when modern methods of purification and instrumental analysis were completely lacking. Nevertheless, I accepted the challenge after making clear my reservations. Asymmetric synthesis could only be hoped for if the monolayer consisted of one or a few two-dimensional single crystals and if the molecules in the two-dimensional crystal to be reacted would all, or for the greater part, be in the same chiral constellation. In practice, this meant that one could forget about a positive outcome within any reasonable time. However, there was an aspect that made it worthwhile to initiate a study on reactions with molecules oriented in a monolayer: such reactions could differ essentially in rate and in product composition from what is seen with conversions in homogeneous solution.

Four years of intensive investigation led to a thesis finished in 1939, shortly before the war reached the Netherlands.[1] The title was "Monolayers, Structure and Chemical Reactions". It is beyond the scope of this chapter to discuss the results in any detail, but a short summary may be in order, because investigations using monomolecular layers or structures built up from monolayers have shown an effective revival.

The electron diffraction technique, inaugurated by Davisson and Germer and introduced at Utrecht by my colleague at the X-ray department, J. de Wael,[2] gave beautiful diffraction patterns with monolayers of compounds such as barium stearate (Figure 1).[2,3] Evidently these films

Figure 1. Monolayer of barium stearate transferred from the water surface onto nitrocellulose, as shown by electron diffraction (1937). (Reprinted with permission from Ref. 3. Copyright 1937 Recl. Trav. Chim. Pays-Bas.)

consist of large two-dimensional single crystals with hexagonal symmetry. With more complex molecules, the monolayers are polycrystalline or semiliquid.[4]

Reactions of molecules in monolayers appeared to have a different rate (and probably also to lead to a different product composition) when compared with similar reactions in homogeneous solution.[1,5]

Substantial amounts of product could be obtained by using a machine for continuous spreading of monolayer, followed by reaction of its components and collection of the product. I enjoyed constructing such a day-and-night-working instrument. The purpose was difficult to explain to the police, who became upset during the night when they saw the fluctuating light from the bulb in the Punch-inspired construction.[1,6]

The war brought this line of research to a stop. The study of a few aspects was taken up again later at Leiden.[7] Of general interest certainly were the results obtained by M. den Hertog-Polak.[8,9] She synthesized and applied "floating" indicator molecules in which the water-soluble pH-sensitive indicator group was balanced by a hydrophobic tail. These dobbers demonstrated visibly that the pH value directly below the surface layer can differ considerably from the bulk pH in the hypophase. Rates of H^+- or OH^--catalyzed processes were found to vary upon introduction of positive or negative charges in the monolayer. In such layers the orientation and the compression of the molecules can easily be changed by varying the surface pressure.

Today's enormously improved analytical methods should make it possible to follow reactions and analyze in detail the products of small areas of monolayer, which would essentially reduce the contamination problem. A continuously operating apparatus can be used for synthetic purposes. With all aspects taken together,[10] reactions with molecules oriented in monolayers can be an attractive tool in the development of organic chemistry and biochemistry (*see*, for instance, the elegant studies of Van Deenen and his colleagues[11]).

In the discussion preceding the study of molecules having a defined orientation as components of a crystalline surface layer, it was considered that the possibility of the spontaneous generation of optically active material is very small but not zero. There is a traditional reluctance to accept the feasibility of systems that, in the course of time, spontaneously transform from inactive to optically active. This reluctance probably has to do with the fact that in fluid media, as a rule, the racemic situation is favored. It occurred to me that this need not be true for systems containing mutually oriented molecules like two- or three-dimensional crystals. In accord with the high symmetry found by electron diffraction for the films used in our investigations, no optical activity was traced with any of the reaction products. However, exemplary systems might be found more easily in the domain of three-dimensional crystallization. There one should search for systems that, without interference from the outside, would have a greater chance of producing a surplus of one enantiomer over the other, rather than staying optically inactive.

This reasoning led to the search for a compound that crystallizes as separate D and L components and racemizes at a rate that is fast in comparison with its crystal growth, and the growth process in its turn is faster than the spontaneous formation of crystal nuclei from a supersaturated solution.[12,13] The essential feature lies in the combination of the three requirements: once a crystal nucleus is formed and starts to grow, the other antipode gets no chance any more because its surplus is being consumed by racemization. The crystallizing antipode may increase to an amount that is much larger than 50% of the racemate originally present.

In the short period available (end of 1938), I found an exciting example by using carefully purified supersaturated solutions of racemic N-allyl-N-ethyl-N-methylanilinium iodide in chloroform.[12,13] The system, after standing for a long time in a closed vessel, yields beautiful clear hemihedric crystals that contain one molecule of crystal chloroform. The example is not ideal, because the rate of racemization is not very fast. Even so, for the most part, only one of the antipodes crystallized (sometimes as a large single crystal) and a great excess of this antipode was formed. The system has an additional nice feature. The optically active product can be dissolved in water, where it dissociates into ions and remains active for months.

With the first series of crystallizations, an instructive complication showed up in that all of the products had the same sign (+) of rotation. This finding was ascribed to the influence of nuclei originating from materials from living organisms. Optically active contents of both signs, as expected on the basis of statistics, were obtained only after extreme precautions had been taken against contamination and possible nuclei. Significantly, crystallization then occurred only after the solutions stood for very long periods at different temperatures.

$$CH_2 = CH-CH_2 \diagdown \underset{N^+}{\overset{\overset{\displaystyle CH_3}{|}}{}} \diagup C_2H_5$$

I$^-$

N-Allyl-N-ethyl-N-methylanilinium iodide.

The latter results seem to demonstrate that the spontaneous formation of optically active compounds from optically inactive material is induced by a statistical process: the formation by chance of a crystal nucleus. Such spontaneous generation of optically active compounds does not play an important role in the development of the powerful methods that aim at the direct synthesis of pure stereoisomers, often containing several chiral centers. However, it remains a scientifically intriguing and significant phenomenon.

In order to round out the introductory part dealing with the scientific activities at the University of Utrecht, it is appropriate to devote some attention to the last part of this period, the years of World War II. To me personally the year 1939 encompassed my promotion to a doctor of chemistry, the happy event of marriage to my partner for life, Louise Oversluys, and the acquisition of a real job in society. I had the (at that time) exceptional fortune to obtain a permanent position as a staff member of the Veterinary Faculty at Utrecht (the same place where, according to Kolbe, Van't Hoff had found his Pegasus). The 5 years spent at the Laboratory of Medical Chemistry of course were not devoted to pure organic chemistry. Moreover, the occupation of our country and the war in general brought intense sadness and catastrophic events that were incompatible with normal life and scientific activity. We tried to continue cultural activities as far as possible.

Valuable cooperation and mutual help existed—as in most other institutions—at the veterinary faculty, a "family" of no more than some 150 students and staff members. I was personally impressed by the professional idealism of my veterinarian colleagues, and I learned a lot of interesting and important physiological chemistry. The director of the laboratory, L. Seekles, passed over the privilege of doing most of the research to his staff members with the unselfish argument that he himself should take care of administration and organization, because he enjoyed a higher salary.

Another very valuable relationship was with H. Veldstra, who had become director of research with the Kininefabriek at Amsterdam. He initiated a stimulating study and joint investigations of the physical

Dinner on the occasion of Havinga's graduation to doctor, January 30, 1939. Behind him paranymphs (assistants to the promovendus) *display his diploma. At his right, L. D. Oversluys, later Mrs. Havinga. Standing between them is J. de Wael, coauthor of Havinga's first two publications. H. Veldstra is to his right and slightly behind him. Behind Miss Oversluys is Havinga's father (his mother was too ill to attend). Across the table are Miss Oversluys' parents. Standing behind Mrs. Oversluys is Dr. H. Erxleben; to her right J. Th. G. Overbeek, who became Professor of Physical Chemistry at Utrecht University. Dr. M. Klompe, the first woman Secretary of State for the Netherlands, is seated at the end of the table, on the right. Professor Kögl stands before the mirror.*

properties and the mode of action of plant growth hormones and analogues of indoleacetic acid. Related investigations on styrene and stilbene derivatives performed later at Leiden University had their roots in these 1942–1946 studies. During that period I was privileged to have C. Schattenkerk as a pupil; she was to become a most gifted co-worker during the following 35 years.

My reminiscences from that war period, on the whole, are truly depressing. However, some things were positive. Mutual help and effort in the preservation of cultural life, science, and art has proven a very valuable asset in the fight against mental erosion.

Finally, in 1945, war and occupation came to an end. People were poor and underfed but, most important of all, free again. The universities started teaching and research with difficulty but with vivid enthusiasm and idealism. For me it meant a new start and a transition to Leiden University.

Ironically, the Leiden spokesman who brought the invitation to

Laboratory building (Veterinary Faculty, Utrecht) where, according to Kolbe, Van't Hoff found his Pegasus.

come to that venerable university was the well-known professor of inorganic chemistry, A. E. van Arkel. During our talks it became clear that we had been in serious competition during the previous years. We both had earned small quantities of food by making at home an adhesive for repairing automobile tires. As unknowing competitors, Van Arkel and I had offered our product to the same repairman. After the war was over, he declared with pride that he had held in his service the complete chemistry faculty of Leiden University!

Stereochemistry: Conformational Analysis

Stereochemistry has been a primary field of study in the Netherlands since the early days of Van't Hoff, who continued the line of thinking of Pasteur, Kekulé, and Wislicenus. Van't Hoff—simultaneously with LeBel—conceived the idea that three-dimensional representations are needed to rationalize the behavior of matter at the molecular level. Concretely, this idea led to the postulate of the tetrahedral carbon atom. More generally, it promoted the acceptance of molecules as real physical entities. It took a long development period before I fully realized how fundamental this concept is to organic chemistry and to major parts of inorganic chemistry, physical chemistry, physics, molecular biology, pharmacy, and many medical disciplines.

This section concentrates on the evolution of what today is called conformational analysis. The important contributions of Sachse and Mohr rightly earned their place in history. The research of Böeseken, Derx, P. H. Hermans, and others at Delft for a long time did not receive the international recognition they deserved. They were some 25 years ahead of their time with what now has to be appreciated as authentic conformational analysis.[14–16]

It is no wonder that stereochemistry, and the study of detailed molecular geometry in particular, was chosen as one of the main fields of research to be cultivated at Leiden after the war. The first to enter research in this direction was G. Dallinga, a senior student who possessed the valuable combination of interest in organic chemistry, physical methodology, and theoretical chemistry.

Six-Membered Rings

We decided to study compounds with nonaromatic ring systems and to make a start with simple cyclohexane derivatives, components of many important natural products. These molecules looked attractive for effective study because we could expect them to show a moderate freedom of conformational change, between the very flexible aliphatic chains and the rigid aromatic and polycyclic systems. Moreover, we enjoyed the stimulus of an advanced (for that time) theoretical treatment by Oosterhoff and Hazebroek.[17,18] They calculated that a chair form would be energetically favored for cyclohexane and its simple derivatives, but that for some other six-membered rings (e.g., 1,4-cyclohexanedione) a flexible form could be equally probable.

J. Böeseken (1868–1948), professor of organic chemistry, Technical University of Delft, 1907–1938.

The determination of molecular dipole moments and electron diffraction were chosen as methods of investigation. The latter looked attractive because in the beginning we liked to obtain data on free molecules in the gas phase. Professor Hassel, whose pioneering studies had demonstrated the possibilities of this method,[19] gave us essential help. Another characteristic of this first endeavor was the organic chemical aspect, which led to a broad choice of molecules to be compared in their properties and behavior.

Dallinga concluded in 1951[20] that *trans*-1,2-dichloro- and *trans*-1,2-dibromocyclohexane in carbon tetrachloride exist as a mixture of two or more molecular forms in dynamic equilibrium. His careful measurements revealed a small but probably significant shift in the values of the dipole moments of these compounds as the temperature changed. He also anticipated that, for a quantitative rationalization of the molecular data, deviations from the regular tetrahedral angles had to be incorporated.

Of course it was not to be expected that definite conclusions on the conformational features of cyclohexane derivatives in solution could be drawn from this first orientation by Dallinga. This became the contribution of Dallinga's successor, W. Kwestroo, who synthesized a representative series of 15 1,2- and 1,4-dihalogenocyclohexanes. Together with F. A. Meijer, Kwestroo measured their dipole moments with solutions in carbon tetrachloride and in benzene.[21,22] We chose these solvents because of a puzzling discrepancy between the value found for the moment of *trans*-1,2-dichlorocyclohexane by Dallinga (2.19 D) and by Tulinskie et al. (2.66 D).[23]

Kwestroo's results (Table I) in the early 1950s were revealing to organic chemists. The dipole moments proved to be the same in the two media for each of the compounds except for the *trans*-1,2-dihalogenocyclohexanes, which showed a considerable solvent effect. Evidently a solvent effect is found with those, and only those, compounds that have different moments in their two chair conformations.

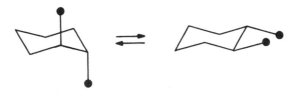

Equilibrium of the two chair conformations of trans-dihalogenocyclohexane: $\mu_{aa} \sim 1\ D$, $\mu_{ee} \sim 3\ D$.

Table I. Dipole Moments
of Dihalogenocyclohexanes
in Benzene and in Carbon Tetrachloride

Substituents	C_6H_6	CCl_4
1,1-Dichloro	2.46	2.48
1,1-Dibromo	2.43	2.45
1,1-Bromochloro	2.47	2.50
cis-1,2-Dichloro	3.13	3.13
cis-1,2-Dibromo	3.15	3.13
cis-1,2-Bromochloro	3.08	3.08
trans-1,2-Dichloro	2.63	2.27
trans-1,2-Dibromo	2.16	1.76
trans-1,2-Bromochloro	2.48	2.04
cis-1,4-Dichloro	2.89	2.89
cis-1,4-Dibromo	2.93	2.89
cis-1,4-Diiodo	2.48	2.45
trans-1,4-Dichloro	0	0
trans-1,4-Dibromo	0	0
trans-1,4-Diiodo	0	0

NOTE: All values are in debyes.

The constancy in the one category and the solvent effect in the other category of compounds led to a convincing description: in solutions of the cyclohexanes, the chair forms strongly predominate. These chair conformations are in a dynamic equilibrium that shifts on changing the solvent. Characteristically, conformers with vicinal equatorial halogen substituents are favored in benzene as a solvent (*see* **Benzene Effect**).

I like to admit that in 1953 this enjoyable capacity to simply rationalize otherwise hardly understandable phenomena made me a believer in the approach now called conformational analysis. Evidently the time was ripe for this development; similar conclusions were reached in the same period by Kozima et al.[24,25] and by Goering et al.[26] At Leiden some 20 students wrote their theses on conformational subjects during the decades following the pioneering investigations of Dallinga and Kwestroo.

Choice of a field of research at the university was often influenced by the desire to offer the students a many-sided program. It became a tradition—traditions are sacred at Leiden University—to compare series of systematically varied compounds by using, without discrimination and in combination, the various methods available for the measurement of molecular properties, preferably in several media. This strategy promotes the observation of novel phenomena, as well as

their interpretation. Curiously, the synthetic effort to obtain the compounds required, difficult as it often proved to be, used to give satisfac-
tion to those who started with a primarily physical or theoretical
interest, and vice versa.

J. A. van der Linden[27] and E. C. Wessels[28] studied the properties
of compounds fixed in one (chair) conformation (*trans*-decalins and *tert*-
butylcyclohexanes). Van Dort[29–31] completed the dihalogenocyclohex-
anes by investigating the 1,3-isomers, and Geise[32–34] made an extension
by tackling the steroid skeleton, one of the basic entities in Barton's
classical treatise.[35] Moreover, the geometry of steroids was of special
interest to us in connection with the vitamin D research. One important
generalization that consistently transpired from the results was the
validity of Dallinga's prediction that valency angles systematically devi-
ate from the ideal value. Molecules of cyclohexane derivatives in the
chair form inherently have a more or less flattened ring with the axial
substituents splayed outward. This feature was established for the first
time in the dioxane series by Altona.[36,37] Steroid skeletons show a substantial degree of curvature (Figure 2). The detailed description and
quantitative data on these phenomena are offered in the theses and
publications cited.

How far confidence and insight grew in the course of many
years of research on conformational equilibria and molecular geometry
can be read from the papers based on the investigations by Hageman.
He started in 1959 with a thorough study of the mechanism and the
steric course of the addition of bromine chloride to cyclohexenes. This
work continued a line of research originating in the synthetic parts of
the studies by Dallinga and Kwestroo and then going to the

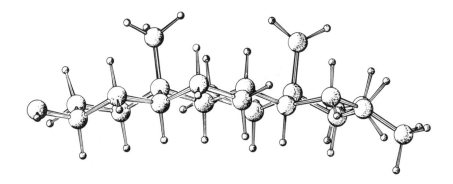

*Figure 2. Curvature of the steroid skeleton (5α-pregnanediol); determined by
X-ray data and VFF calculation.*

investigations of Klapwijk[38] and J. van der Linde,[39,40] who looked into the mechanistic aspects of the mixed addition to double bonds.

Hageman used the occasion to investigate carefully the physical properties and the conformational equilibria of the obtained alkylhalogenocyclohexanes.[41-44] He combined dielectric measurements, NMR, IR, and Raman spectroscopy. The dipole moments of the diaxial conformations of the *trans*-1,2-dihalogeno(alkyl)cyclohexanes proved to be significantly different from zero (~1.1 D). New assignments of the carbon–halogen stretching frequencies were made by Altona and Hageman.[43] Finally a good consistency was reached by the different methods of determining the conformational equilibria of a great number of substituted *trans*-1,2-dihalogenocyclohexanes. Evidently conformational analysis of six-membered carbocyclic compounds had reached a stage of reasonably reliable quantitative description around 1965. Some remaining uncertainties in the vibrational spectra of vicinal ditertiary dihalogenides were clarified in a special study of this class of compounds by Adriaanse.[45]

Benzene Effect

From the beginning of our conformational studies, the benzene effect was regularly used as a test for the occurrence of a dynamic equilibrium in vicinal dihalogenides. However, the benzene effect escaped rationalization on the basis of macroscopic properties like polarity and polarizability. The benzene effect is defined as the change of the free energy difference between *gauche* and *anti* conformers of vicinal dihalogenides and similar compounds on change of solvent from carbon tetrachloride to benzene. A separate investigation was devoted to this effect by Sikkema.[46]

The energy term of the order of 1–2 kJ/mol in favor of the *gauche* conformation in benzene as a solvent is opposed by an entropy term of the order of 1.5 eu. Although it is a small effect, it seems an attractive phenomenon to study in depth, because it opens a way to learn about the subtle specific interactions between molecules that result from their detailed structure and geometry. A benzene effect was found with vicinal dihalogenides, 1,1,2-trihalogenides, 2-halogenoketones, and 3-halogenopropionitriles; with vicinal halohydrins it is absent.

Sikkema postulated as a working hypothesis the formation of an encounter complex in which the two electronegative atoms of the *gauche* conformer closely approach the benzene ring in diagonal orientation. The benzene ring is polarized, and attraction results. In accord with the prediction on the basis of this model, *p*-difluorobenzene as a solvent shows a stronger benzene effect than hexafluorobenzene and methyl-

Benzene effect, as shown by orientation of diequatorial 1,2-dichlorocyclohexane (------) in encounter complex with 1,4-difluorobenzene.

substituted benzenes. There is reason to expect that powerful methods of calculation will lead in the future to a quantitative description of such specific weak interactions between medium-size molecules.

Heterocyclic Compounds: Anomeric Effect

The benzene effect was one of the factors that presented a real surprise in the conformational analysis of heterocyclic compounds. Altona,[36,37] who had undertaken to study the halogenodioxanes, did not observe the expected benzene effect on determining the dipole moments of *trans*-2,3-dichloro- and *trans*-2,3-dibromodioxane (Table II). The values were the same in carbon tetrachloride and in benzene, and between those to be expected for a diaxial and a diequatorial conformation. A critical study using Raman and IR spectroscopy finally led to the conclusion that, in solutions of the *trans*-2,3- and *trans*-2,5-dihalogenodioxanes, there is a strong predominance of just one conformation. This predominant conformation proves to be the same as the one present in the crystal.

To establish the nature of this form with its anomalous dipole moment, Altona took the initiative in what was to become another tradition in the strategy of our conformational-analysis group. With the

Table II. Dipole Moments of *trans*-1,2-Dihalogenocyclohexanes and *trans*-2,3-Dihalogenodioxanes in Carbon Tetrachloride and in Benzene

Compounds	CCl_4	C_6H_6
trans-1,2-Dichlorocyclohexane	2.30	2.63
trans-1,2-Dibromocyclohexane	1.78	2.15
trans-2,3-Dichlorodioxane	1.62	1.63
trans-2,3-Dibromodioxane	1.87	1.90

NOTE: All values are in debyes.

trans-2,3- and trans-2,5-dihalogenodioxanes.

support and guidance of C. Romers, head of the department of diffraction analysis, he made X-ray analyses of his compounds, *trans*-2,3-dichloro- and *trans*-2,3-dibromodioxane, as well as the *trans*-2,5- and the *cis*-2,3-dichloro isomers. It was found that all four compounds occur in the chair form. The *trans* isomers show a diaxial conformation with a considerably flattened ring and the carbon–halogen bonds accordingly splayed outward (Figure 3). A champion in demonstrating the remarkable tendency of the halogen substituents to assume the axial orientation was found in an isomer of tetrachlorodioxane that adopts a tetraaxial form notwithstanding two *syn*-axial repulsions (Figure 4). This tendency was also recognized during research on comparable pyranes, dithianes, and thioxanes. It is similar to the effect observed in carbohydrate chemistry by Edward[47] and studied and termed "anomeric" by Lemieux.[48]

A clue to a valuable description of the phenomenon came from the X-ray data. Altona constructed a device to take Weissenberg diffraction patterns at low temperature, and thus improved on the accuracy of the determination of valency angles and bond lengths.[37] The resulting data revealed that the axial carbon–halogen bond is significantly lengthened in halogenodioxanes, whereas the adjacent carbon–oxygen bond is shortened. In his thesis, Altona put forward the hypothesis that with a C–X–C–Y moiety (X and Y being electronegative atoms like oxygen or nitrogen) in the *gauche* arrangement, there is substantial interaction between the free electron pair on the X atom (the donor) and the antibonding orbital of the C–Y moiety[36,37] (Figure 5). This viewpoint,

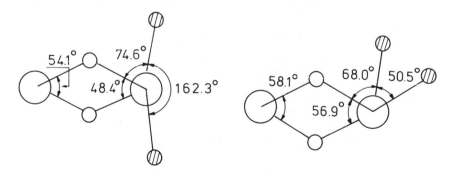

Figure 3. trans-2,3- and cis-2,3-dichloro-1,4-dioxane.

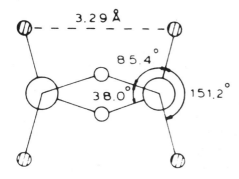

Figure 4.   trans-syn-trans-2,3,5,6-tetrachloro-1,4-dioxane, a champion: four axial substituents (anomeric effect).

based on accurate physical data, finally received additional support from ab initio calculations.

The results by Altona on dioxanes,[36,37] Van Woerden on cyclic sulfites,[49–51] Kalff on dithianes,[52,53] De Wolf on thioxanes,[54,55] and Planje[56] and De Hoog[57,58] on tetrahydropyranes have been the subject of a review article[59] that contains a detailed report complementing the more historically oriented story presented in this chapter. The phenomena found with these six-membered nonaromatic ring compounds could generally be described on the basis of regularities such as the preference for adopting (flattened) chair conformations (which in liquid media are in dynamic equilibrium), the anomeric effect, and the benzene effect. In addition to these general features, each class of compounds has its own specific properties, like the preferred bond angles and the bond lengths about the various heteroatoms. This gives rise to a rich variation of properties and behavior within this class of relatively simple compounds and presents a delight to the organic chemist.

Beyond the domain of the compounds that choose to adopt a chair form, it is historically appropriate to devote attention to the much-discussed 1,4-cyclohexanedione that had occupied a special

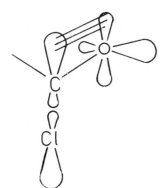

Figure 5. Anomeric effect; stabilization of an axial conformer by delocalization of an electron pair on oxygen through the antibonding orbital of an axial C–Cl bond.

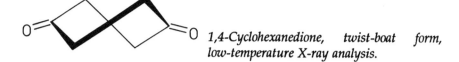

1,4-Cyclohexanedione, twist-boat form, low-temperature X-ray analysis.

position in the early theoretical exploration by Oosterhoff.[17] Mossel made a study of this compound. Again it proved essential to complement the various spectral data by X-ray analysis, performed under the guidance of Romers, and by using diffraction patterns collected at −140 ° C. It was definitely established that this six-membered ring compound, in both the crystalline phase and solution, occurs as a twisted boat with an angle of 155° between the two carbonyl groups.[60–62] In the liquid phase there is rapid interconversion between the two enantiomeric forms.

Seven- and Eight-Membered Rings

More flexibility and variability is to be expected with seven- and eight-membered ring compounds. In 1960 Henniger started to make conformational studies of these systems. At the beginning, results remained rather close to what had been learned from the behavior of six-membered ring compounds. With the *trans*-1,2-dihalogenocyclooctanes, the values of the dipole moments show a benzene effect. This, together with the IR and Raman data, suggest an equilibrium in which a diequatorial boat–chair conformer dominates over a (crown-shaped?) diaxial form and possibly other conformers.

Henniger, in the course of his syntheses, hit on very interesting transannular effects. Following a tradition of the laboratory, he followed up his captivating findings, shifted his attention, and concentrated on mechanistic aspects of reactions of eight-membered ring compounds. The phenomena observed certainly deserved further study.[63–65] The flexibility of the cycloheptane ring was the subject of investigations by Flapper, who applied—along with instrumental methods—molecular mechanics and pseudorotation formalism.[66]

Five-Membered Rings

One of the incentives to start the conformational studies at Leiden by investigating six-membered ring compounds had been the consideration that—besides showing intriguing features in their own right—the six-membered ring forms an essential building block of many classes of molecules in nature. It is not difficult to see that this attractive aspect applies just as well to five-membered ring systems and that the preference for the six-ring molecules in the 1950s came from the feeling that somewhat more was known about them and that they might be less evasive to precise description.

By the 1960s our confidence had grown, and methods of investigation had become much more powerful. It appeared to be a logical step to extend the conformational excursions into the area of molecules with five-membered rings. This field was entered almost automatically with investigations initiated by Geise[32] on the steroids, containing a skeleton of three six-membered rings and one five-membered ring. By bringing this class of compounds to the attention of the conformational people, a profitable bridge was built to vitamin D research, in which at that time the organodynamics had met such fundamental stereochemical phenomena.

The interaction of experiences and insight from the fields of conformational analysis and photochemistry of flexible molecules proved to be very fruitful. Analyses of the X-ray data revealed that the five-membered ring shows a nonplanar conformation that varies somewhat in the different steroids. A later thorough review[67] mentions the following as general features:

1. With the fused polycyclic skeletons there is substantial variation of bond angles and even bond lengths about the standard value; C–C bonds individually may adopt values from 1.50 to 1.58 Å in minimizing the total molecular energy.

2. Steroids with 10β and 13β methyl groups generally show the characteristic curvature established by Geise.[32]

3. The five-membered D ring is nonplanar; in each compound it adopts a specific geometry that conforms to one of the positions on the pseudorotation circuit first formulated by Pitzer and Donath for cyclopentanes[68,69] (Chart I).

It was very fortunate that the investigation of low-molecular-weight halogenopentanes, tetrahydrofurans, and indanes fell to the able brain and hands of H. R. Buys, who in a remarkably short time found his way in the phenomena exhibited by the flexible molecules. For his thesis research[70] (1965–1968) he exploited the by then time-honored combination of methods: synthesis of the selected variety of compounds; determination of dipole moments; and IR, Raman, and NMR spectroscopy. One of his original approaches was the use of a linear relationship between the square of the dipole moments μ and the vicinal coupling constants J within a series of isogeometric molecules:[71]

$$\frac{d\mu^2}{dJ} = \frac{\mu_{EE}^2 - \mu_{AA}^2}{J_{EE} - J_{AA}} \tag{1}$$

This relation enables us to check the consistency of the results and to estimate the value of a dipole moment or a coupling constant if one of the properties in a series cannot be measured.

References 70 and 71 provide a review of the wealth of interesting data and phenomena found concerning the molecules with a five-membered ring. In substituted cyclopentanes and comparable compounds, there often is restricted pseudorotation (pseudolibration) about one or two favorable conformations. Thus, in the discussion, we speak of substituents in (quasi)axial or equatorial orientation and of envelope forms and half-chair forms (Chart II). The benzene effect shows up analogously to what has been discussed for the six-membered ring compounds. Keep in mind, however, that the energy minima and maxima in the pseudorotation circuit are inherently smaller than with most six-membered rings with their favored chair conformations. In fact, with the five-membered rings, the substitution pattern determines the preferred conformations. For this reason we cannot use the well-known "holding-group" approach.

In the 1970s and 1980s, the conformational considerations of polynucleotides would be based on the insights gained with simple tetrahydrofurans and applied to the furanosides of DNA and RNA.

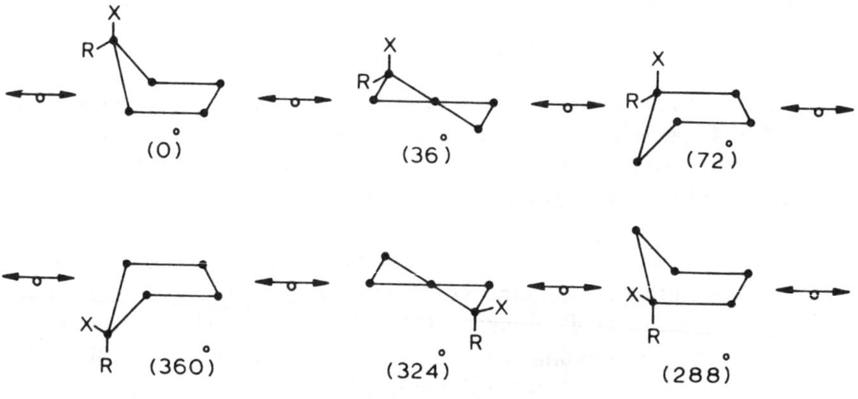

Chart I. Pseudorotation in cyclopentanes. Two parts of the pseudorotation circuit are shown for 1-halogeno-1-alkylcyclopentane: halogen axial (above) and equatorial (below). The numbers in parentheses indicate the values of the phase angles in the pseudorotation circuit.[68,69]

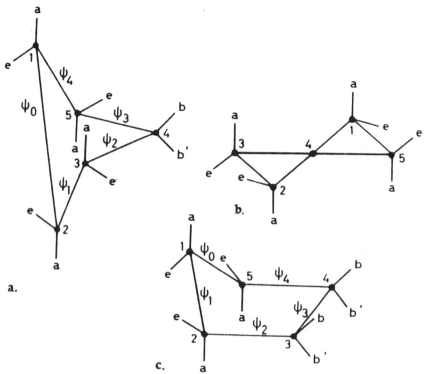

Chart II. Cyclopentane: half-chair form in perspective (a), projection (b), and envelope form (c), with axial, equatorial, and bisectional valencies.

Developments After 1970

As in many other fields of science, the rapid growth of computational facilities and methods led to new vistas in stereochemistry. Altona, who developed into an excellent research leader and in the 1970s became a professor of chemistry at our institute, together with Hirschmann initiated the application of molecular mechanics to the conformational analysis of steroidal molecules.[72–74] The Valence Force Field (VFF) calculations had considerable success, above all in being able to accurately reproduce the curvature of the steroid skeleton as deduced from X-ray data by Geise.[32] A further interesting feature was that ring B of Δ^5- and $\Delta^{5,7}$- steroids may function as a hinge in the molecule. This may be significant in biological interactions.[67]

The most impressive development occurred with the various NMR methods. It is illuminating to realize that only 17 years ago Lugtenburg in his thesis research introduced the application of the nuclear Overhauser effect (NOE) in the Netherlands.[75,76] The powerful new methods with their multidimensional representations, their cosyness, and other nice relationships, came into general use during the last decade. Powerful (500-MHz) machines and apparatus for a variety of nuclei have become generally available. Organic chemists, the students of detailed molecular structure and behavior par excellence, have to be experts in the application of these physical tools to scrutinize their objects of study. The latter range from the dynamics of small and medium-size molecules to the conformational behavior and reactions of macromolecules. As an example of the current state of the art, Figure 6 shows a picture of a novel "hairpin" octanucleotide, for which the deduction of the complete conformation appeared feasible.[77] This geometry, with the characteristic N(orth) and S(outh) furanoside rings, is based on the fundamental studies mentioned in connection with the five-membered ring compounds.

Although a detailed discussion of the methods used cannot be the aim of this contribution, I cannot help but stress the importance of the well-known Karplus relation between NMR coupling and geometry:[78]

$$^3J_{HH} = A \cos^2 \phi + B \cos \phi + C \qquad (2)$$

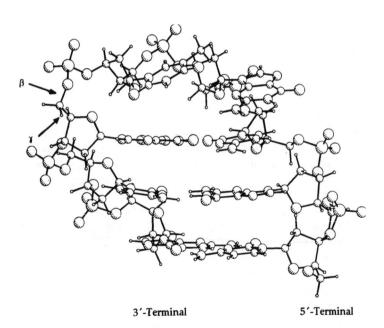

3'-Terminal   5'-Terminal

Figure 6. Conformation of the hairpin form of a DNA octanucleotide, featuring a loop of only two nucleotides.

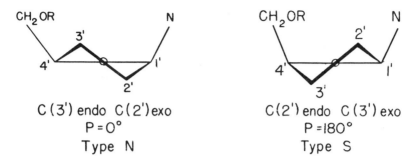

N *and* S *conformations of ribose ring in polynucleotides.*

This equation has been (mis)used beyond its limits, as warned against by Karplus himself. But even so, it has been stimulating to the extent that the positive influence greatly overcompensates for possible mistakes caused by stretching the application too far. Moreover, it has been possible to formulate more complicated but still tractable extensions, like the one deduced by Haasnoot, De Leeuw, and Altona.[79] Such relationships will remain a valuable asset to conformational analysis in years to come. It seems probable that in the near future adequate computers will enable the determination of large areas of energy hypersurfaces of molecules and the description of molecular behavior beyond the areas around the energy minima. This then may bring conformational analysis to the stage where it can indicate the routes followed by reacting molecular systems. Signs foreshadowing this evolution of organodynamics can be seen in the approaches by Karplus and other chemists on the dynamic behavior of (biological) macromolecules.[80]

Additional Remarks

This section concentrates on developments as they took place with the Leiden group. Of course there have been fruitful interactions and overlap with the investigations at other places. The latter will be dealt with partly in other volumes of this series. A more complete presentation is offered in a few excellent texts like those of Eliel[81,82] and of Hanack,[83] in review articles, and in the series *Topics in Stereochemistry*. I would like to pay tribute here to those scientists who, by writing and editing such review texts, made invaluable contributions to the benefit of their colleagues and to the development of stereochemistry. I would also like to express my gratitude for the exchange of ideas and experiences over time with colleagues studying conformational analysis and related fields.

As a European chemist, one cannot deal with developments of stereochemistry without mentioning the Bürgenstock conferences on stereochemistry. These yearly meetings, although slightly different from the renowned Gordon conferences with respect to the housing facilities, had a comparably stimulating influence. The conferences, initiated by A. Dreiding and a group of his colleagues from Zürich, cultivate some admirable traditions. One of these states—probably correctly—that practically all of interesting novel chemistry can be considered part of stereochemistry. Another tradition rules that every year at least one presentation should be devoted to highly fantastic chemistry.

You may have noticed that no definition of "conformation" has been given. This again has to do with a literally breathtaking nonstop evening discussion on this subject. Thanks to the presence of the highest authorities in the field, the impossibility of reaching an accepted definition was firmly established. But even without these unique aspects, it would remain strongly desirable for the development of chemistry in Europe that the Bürgenstock conferences continue to flourish for a long time in the future.

1974 Bürgenstock Conference, on Havinga's 65th birthday. The conference is held annually during the end of April–beginning of May. Havinga, whose birthday was May 7th, often celebrated the occasion during the conference. Left to right: The late H. Kloosterziel, A. van der Gen, Havinga, J. D. Dunitz, W.N. Speckamp, K. Schaffner, E. L. Eliel, C. Altona.

The 1984 Bürgenstock Conference. R. H. Martin and Havinga propose a toast to V. Prelog. Joining them are Mrs. R. Scheffold, Mrs. L. Ghosez (seated to Martin's right), Wang Yu, A. Dreiding, and Mrs. Dreiding.

Vitamin D: Photochemistry and Conformational Behavior

In retrospect, it now seems logical that specific effects of orbital symmetry, conformational equilibrium, and electronic arrangement in molecular reactions were to be traced by comparing the conversions of molecules with identical chemical constitution but with different electronic escort [i.e., by comparison of reactions of molecules in the ground state (thermal reactions) and the same molecules brought into an excited state (photoreactions)]. It was a bizarre coincidence that at Leiden in the early 1950s we wandered into two fields where the comparison of photoinduced and thermal processes presented examples of a then-novel phenomenon: the oppositeness in stereospecificity and regioselectivity between photoreactions and ground-state reactions.

The first field was vitamin D chemistry, where the opposite stereospecificities of the photoinduced ring closure and the thermal ring closure were established with a conjugated Z-triene, previtamin D. The second field consisted of the nucleophilic aromatic substitutions, in which *meta* (!) activation was discovered with the photoreactions in contradistinction to the orthodox *ortho–para* activation in the thermal substitutions.

The detailed exploration of each of these fields in the course of the ensuing 30 years called for considerable attention and energy at the Leiden laboratory. Although it was inferred before long[84] that a similar rationalization would prove adequate in both areas, the investigations of the vitamin D group and the photoaromatic group developed along their own pathways. It was only during the last decade that the lines of research started to approach each other and arrived at a related description. For that reason, and in order not to make the stories too complex, developments in the two fields will be discussed separately, starting with vitamin D and the conjugated dienes and trienes in general.

Vitamin D and Related Polyenes (1950–1965): (Photo)Reversible Reactions

The first contributions to vitamin D chemistry in the Netherlands were made by E. H. Reerink and his colleagues in the late 1920s. Their photochemical research at the Philips laboratories (Eindhoven), isolation of

E. H. Reerink, originator of research and industrial production of vitamin D in the Netherlands.

Ergosterol (E) $\xrightarrow{\text{hv}}$ Lumisterol (L) $\xrightarrow{\text{hv}}$

Tachysterol (T) $\xrightarrow{\text{hv}}$ Vitamin D (D)

Scheme I. Windaus' scheme of the photoinduced conversion of ergosterol into vitamin D.

the antirachitic vitamin in the form of mixed crystals from irradiation mixtures of ergosterol,[85] placed them among the pioneers.

A flourishing industry developed as a direct result of Reerink's scientific accomplishments. He then gave effective long-term guidance to this industry as a director. Reerink, Van Wijk, and Van Niekerk concluded from their meticulous spectroscopic investigations in 1932 that there must be a precursor that transforms into vitamin D[86] through a reversible thermal reaction when the temperature is not too low. The isolation of this precursor, previtamin D, was accomplished in 1948 by the research of Velluz and his colleagues.[87] Its chemical structure was determined practically simultaneously by our group and by the French authors in 1955.[88,89]

At that time there was a generally accepted reaction scheme for the photoinduced formation of vitamin D from ergosterol (and 7-dehydrocholesterol). It consisted of three consecutive photochemical steps, with the compounds called lumisterol (L) and tachysterol (T) as the first and second intermediates between ergosterol (E) and vitamin D (D) (Scheme I). The scheme had been proposed by A. Windaus. In a classical series of investigations with his co-workers in the 1930s, he accomplished the structure determination of several of the isomers involved in the process. As some authors, (including my former fellow student, J. van der Vliet) pointed out, there were reasons to doubt this scheme.[90,91] However, probably because of the authority of Windaus, it survived for more than 20 years.

At Leiden we chose vitamin D as one of our research subjects, because we were inspired by the knowledge and the results in this field obtained in our country (Philips, Eindhoven–Weesp) in the years 1925–1945. Our first efforts were aimed at unraveling the metabolic pathways followed by vitamin D upon oral and intravenous administration. To this end, Bots synthesized the labeled provitamin D_3, 3-^{14}C-7-dehydrocholesterol, and the labeled vitamin D.[92] Bots and then Van den Bos obtained some interesting results with C-26- and C-27-labeled compounds.[93,94] They reached the conclusion that vitamin D is rapidly metabolized throughout the body, not only in the intestines. However, techniques at that time were not sufficiently advanced to make detailed metabolic studies with the vitamin. This was accomplished some 20 years later by the important studies of Kodicek, DeLuca, and Norman.[95-97]

Once labeled provitamin D was available, it occurred to us that this could be used to trace the pathways of photoconversions in the vitamin D field (e.g., to investigate the Windaus scheme or alternative schemes). Two senior students, A. L. Koevoet and A. Verloop, in 1952 enthusiastically embarked on the project in a close and effective cooperation. We decided to orient ourselves rather broadly in vitamin D chemistry and to study phenomena by a variety of chemical and physical methods.

An informative result, reported in one of the first publications of Koevoet and Verloop, gave us a flying start.[88] The structure and configuration of previtamin D could be established on the basis of its UV and IR absorption, its rate of reaction in Diels–Alder condensations, and finally by its smooth isomerization to tachysterol under the influence of iodine and light. Evidently previtamin D is the Z isomer of tachysterol, for which the E configuration had been deduced by Inhoffen, Bruckner, Grundel, and Quinkert[98] at Braunschweig. Velluz, Amiard, and Goffinet arrived independently at the same view.[89] Chart III presents the formulae of the five isomers, with the configuration of lumisterol as deduced by Castells, Jones, Williams, and Meakins at Oxford.[99]

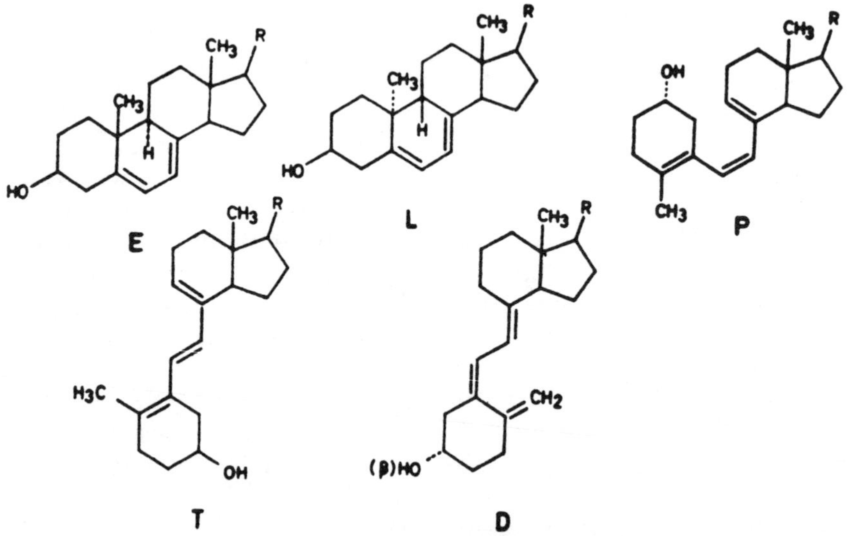

Chart III. Structural formulae of isomers occurring in the formation of vitamin D from ergosterol. Key: D, vitamin D; E, ergosterol; L, lumisterol; P, previtamin D; T, tachysterol.

Another early conclusion, established by ^{14}C tracer experiments, was that neither lumisterol nor tachysterol is an obligatory intermediate in the transformation of provitamin D (7-dehydrocholesterol) to vitamin D.[100–105] Therefore, the Windaus scheme had to be abandoned.

The reversibility of at least some of the reactions[89] made it very difficult to prove or disprove other possible schemes that were proposed by several authors. The essential correctness of a scheme suggested by Velluz[89] was finally established by G. M. Sanders.[106] A rigorous analytical exploration appeared essential, and this was undertaken by Rappoldt.[102,105] Systematically, starting with a solution of one of the isomers E, L, T, and P, Rappoldt determined the composition of the irradiation mixture after different (short) times of illumination. Extrapolation to zero time then gave the primary photoproducts of the isomer under consideration, qualitatively and quantitatively. Quantum yields of each of the pathways of photoproduct formation were measured. Rappoldt's inspiring investigations, published in 1958–1960, covered the majority of the reversible reactions around previtamin D.[107] The network, completed by G. M. Sanders,[106,108] led to the scheme of reactions with their quantum yields as presented in Scheme II.

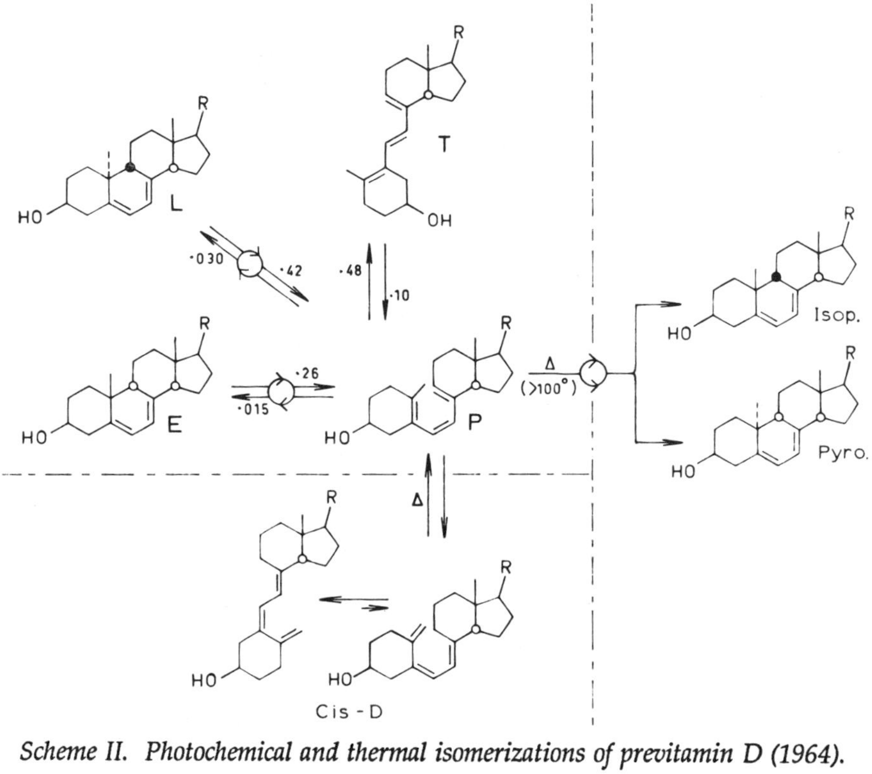

Scheme II. Photochemical and thermal isomerizations of previtamin D (1964).

Mechanistic Studies: Orbital Symmetry and Other Factors

Even before we had the complete scheme finished, and while we were still hampered by ignorance of the pathway of formation of lumisterol, we could not resist doing mechanistic studies and making some speculations. Investigations into systems at very low temperature (fluorescence, no phosphorescence) made it clear that photoisomerizations of the vitamin D isomers start from the first excited $\pi-\pi^*$ singlet state.[107,109] The shortness of the singlet lifetime was one of the factors that induced us to introduce[110,111] what later was termed the NEER (nonequilibration of excited rotamers) principle.

Scheme II reveals some historically interesting processes. There is an amazingly smooth [1,7] H-shift in the thermal equilibration between previtamin D and vitamin D (Figure 7). For this shift, found also with model compounds, an antarafacial pathway was deduced on steric grounds in 1961 before we arrived at the concept of orbital symmetry as one of the factors determining the course of multicenter reactions.[110,112]

The vitamin D–previtamin D reaction is interesting not only for its efficient mechanism, but also for the remarkable position of its equilibrium. It is 80/20 in favor of the vitamin, which seems efficient for functioning in biological systems. However, it contrasts with the behavior of model compounds as synthesized by Schlatmann[112,113] and with theoretical expectation that disfavors situations with double bonds exocyclic to the cyclohexane ring. A satisfactory explanation was offered by the investigations of Pot[114] on the 13α isomeric compounds and of Takken[115,116] on analogues with various sizes of the D ring. The results presented in Table III are heartwarming to the organic chemist. Subtle variation of molecular structure directly shows that the equilibrium shifts as a function of the internal strain imposed on the cyclohexenoid C ring and the D ring in the previtamin-D-type molecules. This strain depends on the mode of the ring junction and on the size of the D ring in a rational way (Figure 8).

Scheme II further shows the photoinduced ring closures of the 6π-electron system that specifically transform the Z-triene previtamin D into the 9,10-*anti*-cyclohexadienes E and L in contradistinction to the

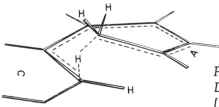

Figure 7. Previtamin D–vitamin D transition state; antarafacial [1,7] H-shift (1961).

Table III. Previtamin D—Vitamin D Equilibrium
as a Function of the Size of Ring D and Its *trans* or *cis*
Connection to Ring C

P	D	%P	%D
		>99	<1
		>95	<5
		65	35
		10	90
		0	100

Figure 8. Newman projection along C-14–C-13; opposing influences of D-ring and C-8–C-9 double bond give rise to strain in previtamin D.

thermal "disrotatory" cyclizations that yield the 9,10-*syn* isomers pyro- and isopyrocalciferol.

This striking contrast stirred the imagination of the members of the vitamin D group. Our guide in theoretical problems, L. J. Ooster- hoff, and the *promovendi*, J. L. M. A. Schlatmann and J. Pot in particular, made essential contributions in the discussions toward the view that— next to steric and trajectory factors (rotation barriers)—the symmetry of the highest occupied molecular orbital determines the pathway of such a "multicenter" reaction.[110,111] Curiously this 6π cyclization has been quoted in the literature as the case in which orbital symmetry was indi- cated for the first time as a reaction-influencing factor. However, this effect of orbital symmetry is invoked earlier in the text of the same 1961 publication, in the description of the photoinduced 4π ring closures of pyro- and isopyrocalciferol. From the beginning in 1961, orbital sym- metry was considered as a general influence and presented for the buta- diene as well as for the hexatriene cyclization.[110] It was complemented by "trajectory considerations" (preferred modes of rotation in the excited state as compared with the ground state).

Expositions of the experimental data and the rationalizations were also given in various lectures at congresses in Europe and in con- siderable detail at the 1963 photochemistry congress at Rochester. Grad- ually the results and considerations of the Leiden group became known and quoted. Then, as is well known, the development was greatly accelerated through the expositions of Hoffmann and Woodward, who encountered the oppositeness of thermal and photoinduced cyclizations in the course of studies on the vitamin B_{12} synthesis, again with the hexatriene–cyclohexadiene system. In their first publication (1965), they offered the same explanation based on the symmetry of the highest occupied orbital.[117] Soon followed the series of brilliant publications of increasing generality and thoroughness by Woodward and Hoffmann in the first place, by Longuett-Higgins and Abrahamson,[118] Fukui,[119] Dewar,[120] Zimmerman,[121] and others.

In the years before these epoch-making publications, the rational- izations given by the Leiden group spread relatively slowly. The main reason probably was that the novel concepts were published compactly

Luitzen J. Oosterhoff (1907–1974), professor of theoretical organic chemistry, University of Leiden, 1950–1974.

formulated, in the context of research on vitamin D and organic photochemistry, which at that time was in its infancy. Moreover, the experimental reaction scheme was not completed before 1964.[106] It was felt at the theoretical front that the orbital symmetry argument needed essential improvement, particularly for the description of the photochemical processes. The latter processes as a rule proceed partly in the electronically excited state and partly in the ground state.

Here I wish to express my admiration for the late L. J. Oosterhoff who, in the 1960–1961 discussions, was the first to suggest orbital symmetry as a reaction-influencing factor. At a later stage (with his co-workers Van der Lugt, Mulder, and Van der Hart),[122–124] Oosterhoff gave the first satisfactory and quantitative description of the pathway and driving force of the photoinduced conversions. As a tribute to this great scientist and admirable man, the 1968 diagram for the butadiene cyclization is reproduced with its exemplary energy curves and the "Oosterhoff minimum" through which the ground state is reached and the photoproduct is formed that corresponds to a high-energy barrier in the thermal conversion (Figure 9).

Conformational Equilibrium and Photoproduct Composition: NEER Principle

Another general concept introduced in the explanatory 1961–1962 publications[110,111] concerns the relationship between conformational equi-

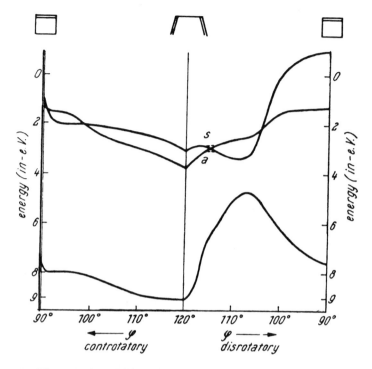

Figure 9. The energies of the ground state and two excited states during the butadiene–cyclobutene reaction. (Van der Lugt and Oosterhoff[122])

librium (of the conjugated trienes and dienes in the ground state) and photoproduct composition. These considerations originated from our research in conformational analysis, which brought with it the exploration of molecular geometry changes as a function of the magnitude of rotation barriers, and from the experimental finding that the reacting excited species in the direct photoconversions of trienes and dienes is in a singlet state.[107]

On the basis of the short lifetime of the π–π^* singlet, combined with the relatively high double bond character of those bonds of the conjugated system that were single before excitation (Scheme III), it was inferred that the excited conformers would not have the opportunity in their short life to reach thermal equilibrium. In brief, ground-state conformers behave as configurational (E–Z) isomers being promoted to the π–π^* singlet. Each conformer, after absorption of light, gives its own specific mixture of products (Figure 10).[116] In a way, it represents the counterpart of the Curtin–Hammett principle that is characteristic of thermal processes.[125]

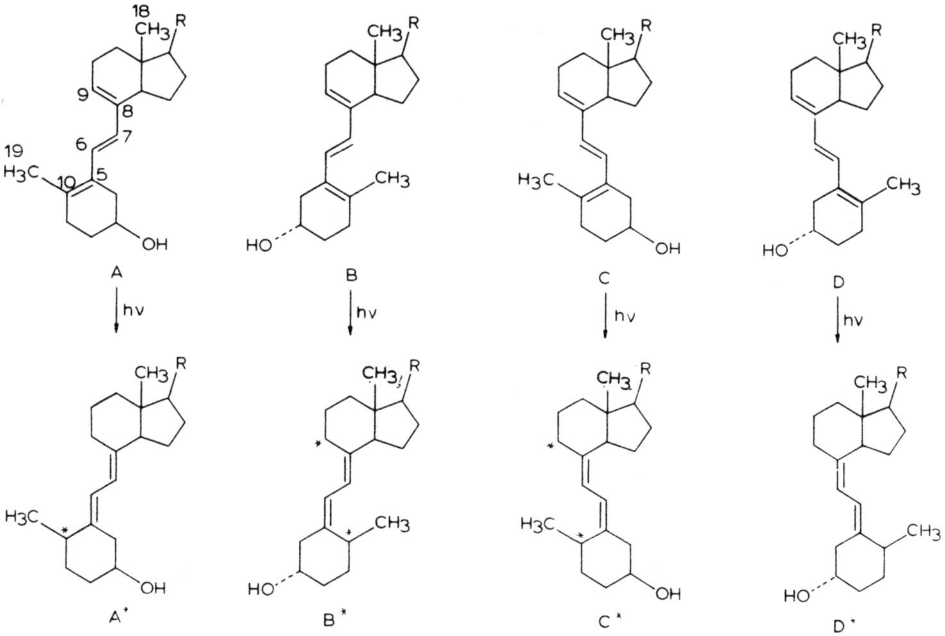

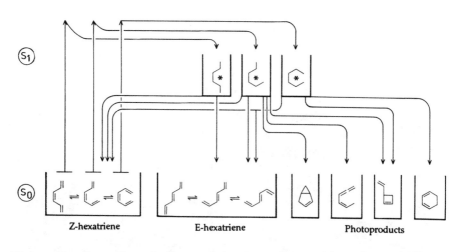

Scheme III. Schematic representation of conformations of the ground state of tachysterol and of the corresponding $\pi-\pi^*$ excited states.

Figure 10. Formation of photoproducts from three conformations of (Z)-hexatriene according to the NEER principle[116]. (Reprinted with permission from Ref. 116. Copyright 1973 Experientia.)

The concept of the nonequilibration of excited rotamers was used in the early publications of vitamin D photochemistry to rationalize the efficiencies of formation of the various photoproducts from tachysterol and precalciferol.[110,111] Colleagues in physics, who "grew up" with the Kasha rule for fluorescence, often considered it absurd that such energy-loaded particles would obey these restrictions. I was pleased to learn that Hammond and Liu in 1963 had courageously and successfully applied a similar reasoning to rationalize the product composition of the reactions of triplet-sensitized butadienes,[126] even though with triplets the "natural" lifetime of the excited conformers may be considerably longer. Moreover, results of further studies (e.g., those of Dauben,[127,128] of John E. Baldwin,[129] of Courtot,[130] and of the Leiden group[131,132]) strengthened belief in the validity of the principle.[116]

The NEER (nonequilibration of excited rotamers) principle was finally proven by the investigations of Gielen, who in simple systems (2,5-dimethyl-E-hexatriene and 2,5-dimethyl-Z-hexatriene) established a striking difference in product composition at very low degrees of conversion upon irradiation with light of different wavelengths.[133] We thereafter found it justified to introduce the term NEER into the literature.[134]

In subsequent experiments, it could be specified that such a wavelength effect—existing by virtue of the NEER concept—of course

Discussion session of the vitamin D research group in Havinga's office, 1978. Left to right: J. W. J. Gielen, R. B. Koolstra, Havinga, S. J. Halkes, G. Lodder.

results from the difference in absorption spectra of the various rotamers. In addition, it may be caused partly by a difference in product composition of one and the same conformer, when illuminated with light of different wavelength.

A very elegant proof of the reality of the NEER principle came from spectroscopic disciplines. E. Fischer, in a series of investigations with substituted stilbenes and related compounds, could directly bring *ad oculos* the various conformers of the ground-state equilibrium by analysis of the luminescence spectra upon shifting the wavelength of excitation.[135]

Recently, it has been deduced from time-resolved resonance Raman spectra that the NEER principle, as anticipated by Hammond and Liu[126], can be extended even to excited triplet states.[136]

Vitamin D, Dienes, and Trienes (1965—1980): Irreversible Photoreactions, Suprasterols, and Toxisterols

In 1965 the investigations of G. M. Sanders had rounded off the research on the products and stoichiometry of the "reversible" photoreactions in the vitamin D field.[137] Moreover, an effort to rationalize the processes had been made by introducing as directive factors the influence of the orbital symmetry and of the conformational composition of the starting compound in the ground state. It therefore seemed a logical follow-up to start the exploration of what had remained a void in vitamin D chemistry: the field of "overirradiation products".

These compounds, described rather vaguely in the literature, are the products remaining after long-term irradiation of vitamin D or its precursors. Evidently they are formed by monodirectional photoreactions: suprasterols from vitamin D itself and toxisterols from previtamin D and tachysterol. I was encouraged to investigate this field by W. G. Dauben, who himself had determined at an early date— among many other important contributions to vitamin D chemistry—the structural formulae of the two most abundant suprasterols, SI and SII.[138,139] These are photoisomers of vitamin D, in which the hexatriene moiety is transformed into a bicyclo[3.1.0]hexene construction (structures, Scheme IV).

Efforts to isolate and identify more irradiation products of vitamin D received essential help from the results obtained in the parallel study of methyl-substituted hexatrienes performed by Vroegop.[131,132] His experiments with these simplest possible hexatrienes very clearly demonstrated the usefulness of considering the ground-state conformational equilibrium as a factor determining the photoproduct composition. With vitamin D itself, Bakker and Lugtenburg[140] traced exactly

suprasterol I

suprasterol II

Suprasterols SI and SII.

those products that were expected from consideration of the two favored conformations known from optical and X-ray data. Isolation and structure determination were accomplished by short-column chromatography, chemical conversions, and spectroscopic methods (UV, IR, and NMR in particular).

The complete vitamin D photogenetic family is presented in Scheme IV. With the allenes (SIII and SIV), interconversion could be effected by irradiation in their absorption band; their diastereoisomerization corresponds to the predictions made by Van't Hoff in 1875. Such allene derivatives have been used by Okamura[141] as intermediates in effective syntheses of vitamin D derivatives, because they smoothly isomerize into the corresponding hexatrienes at relatively low temperature.

In the field of the photoproducts whose ominous name, toxisterols, appears to be unnecessarily terrifying, we were strengthened by the study of Westerhof and Keverling Buisman,[142] who isolated both toxisterol A and toxisterol B. Both compounds show the absorption at around 250 nm that was reported in the literature as characteristic for toxisterols. The indication that toxisterol B represents an adduct with one molecule of solvent (alcohol) was supported by the investigation of Sanders,[108,143] but overall information on toxisterols remained rather vague and fragmentary.

The story of the disclosure of this field, with the isolation and structure determination of no less than 11 toxisterols, demonstrates that, even with powerful modern techniques, it takes the perseverance and ingenuity of the chemist to unravel really delicate systems. Boomsma, who accepted this challenge as the test for obtaining the doctorate,

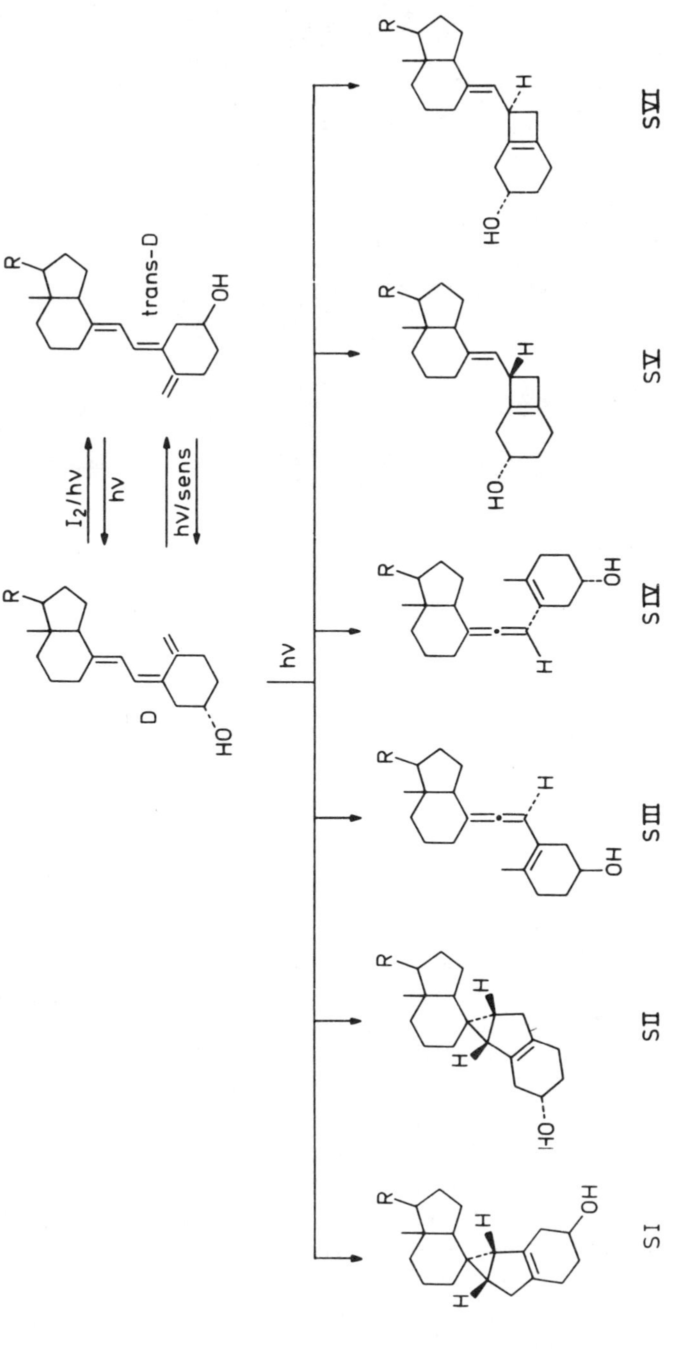

Scheme IV. *Photoproducts of vitamin D (1972).*

Berlin, 1979. Fourth Workshop on Vitamin D. Left to right: J. Lugtenburg, Mrs. H. L. Henry, W. B. Whalley, unknown, W. G. Salmond, Mrs. Havinga, A. W. Norman, Havinga, W. H. Okamura, T. Suda, unknown, J. W. J. Gielen, H. J. C. Jacobs.

spent 2-1/2 years on perfecting methods of chromatographic separation and isolation. He then achieved the well-earned success of disclosing practically the whole toxisterol field within less than a year.[144-146]

Scheme V presents a survey of the results, which partly conform to reactions predictable on the basis of existing knowledge and insight (toxisterols C, D, and E). On the other hand, some compounds are formed by conversions that are novel to hexatriene photochemistry (toxisterols A and B). Barton and his collaborators also found and studied toxisterols A and B and isolated two intramolecular ethers formed with low quantum yield.[147] There may well be a few other irradiation products formed in very low yield, but the material balance is satisfactory if we count the products presented in Scheme V.

The consideration of possible pathways for the formation of the many new compounds represented an exciting enterprise that was tackled at Leiden by H. J. C. Jacobs, an eminent senior co-worker in the vitamin D field since the early 1970s. Along with the factors mentioned in earlier rationalizations, an additional point of view that appeals to the organic chemist may be useful in describing the formation pathway of some classes of photoproducts. We can assume that one of the relaxation pathways of excited conjugated trienes starts by rotation about the central bond to yield two allylic halves. These allylic species may react

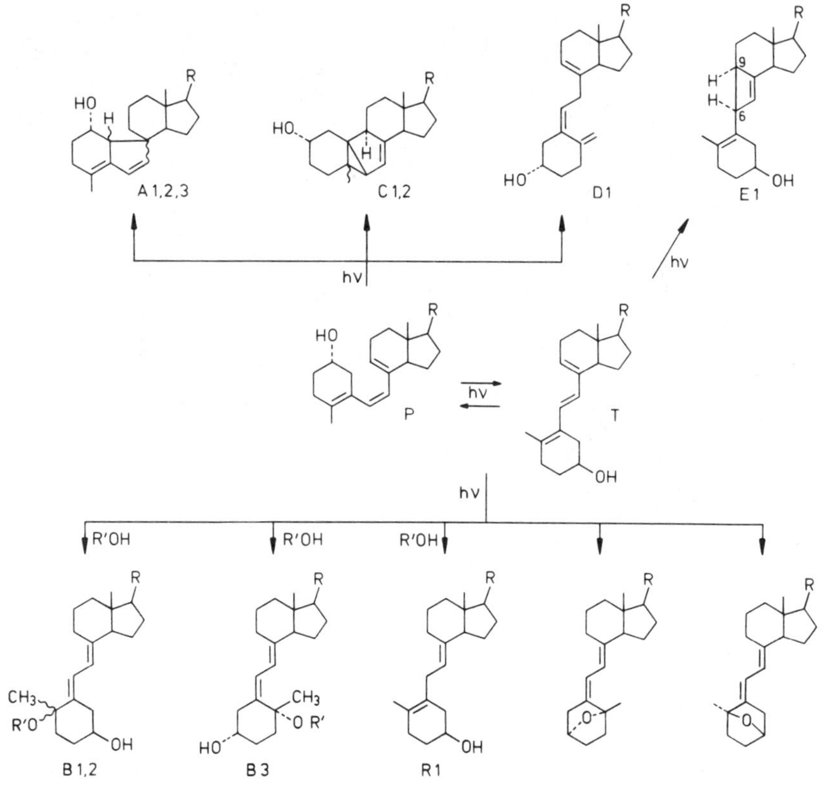

Scheme V. "Overirradiation" products of previtamin D–tachysterol: toxisterols (1975).

as radicaloids, or they may assume an ionic character, one half becoming positive during the reaction and the other half becoming negative. For reactions with a reasonable quantum yield, I prefer to indicate the tendency to react via a (crypto)ionic mechanism as derived from a high polarizability. This mechanism implies a less strict angle dependency than that required by the highly interesting and important concept of sudden polarization.[148]

Even so, the usefulness of this description has to be checked by studying the influence of the polarity of the reaction medium, of electron-donating and -attracting substituents at the conjugated system, and of the nature of the reaction partner in bimolecular processes. Tentative investigations in this direction suggest that simple guidelines will not be established easily.[149,150] A look at the richness of phenomena and products in the toxisterol field warns us that we should not expect to arrive at a satisfactory description without going into the details of the

Berlin, 1979. Fourth Workshop on Vitamin D. Left to right: J. Lugtenburg, E. Zbiral, M. P. Rappoldt, Havinga, H. J. C. Jacobs.

structures of the reacting molecules and their interactions with the solvent. On the other hand, results like those obtained by Maessen and Jacobs[149] mark the start of promising excursions exploring the wealth of photoinduced reactions of polyconjugated systems, of which previtamin D is a shining example. It seems appropriate to indicate the situation by mentioning some results that can be transmitted in a few words.

The nature of the solvent and the position and character of substituents significantly influence the conformational equilibrium and thereby the photoproduct composition. This influence manifests itself more clearly than that stemming from regulation of the polarizability of the polyene system.[149,150]

Use of vitamin D with its two conformations of similar UV absorption led to indications that an excited rotamer, guided by the NEER principle, at least in certain cases has a choice between different corridors of relaxation. The different relaxation pathways may result in different photoproduct compositions for one and the same conformer, depending on the wavelength of irradiation.[151,152]

Remarkable phenomena were observed upon irradiation of the Z-triene previtamin D at about 90 K. It looks as if one of the unstable conformations of the E triene (tachysterol) can be seen and studied as long as the temperature remains low enough.[153]

The most recent developments (Brouwer and Jacobs[154]) are in the study of the simple methyl-substituted hexatrienes. These investigations aim at exploring in detail the reaction trajectories followed by the molecules upon excitation. This study will call for a combination of the use of isotopically labeled compounds, of the many optical methods now available (including the ultrashort-pulse techniques), and of the calculation of conformational relationships beyond the areas of energy minima. Such endeavors finally may result in maps of the energy hypersurfaces of the excited states and the ground state. Combination of the results obtained with the simple molecules and with the stereochemically informative natural products should lead to a basic understanding and rational application of photochemical processes.

Photochemistry: Aromatic Photosubstitution

Whereas most investigations I have discussed resulted from planned research in a chosen field, investigations at Leiden on photochemical processes—besides the reactions in the vitamin D field—originated from a chance observation made in the course of experiments that had a totally different background. In the early 1950s, we decided to study the reactivities of various nitrophenyl phosphates in order to learn about the chemical mechanism of decoupling of energy transfer in biological systems, as shown in an exemplary way by 2,4-dinitrophenol.

R. O. de Jongh, a senior student who started the research on this subject, had no trouble in determining accurate rate constants for the very slow alkaline hydrolysis of the phosphates of *o*- and *p*-nitrophenol. He became annoyed that he could not get consistent results with the even less reactive *meta* compound. Sometimes it looked as if the reaction of this isomer tended to slow down during the night. Revealing light was shed on this disturbing feature one day when De Jongh observed that a solution of the *meta* compound to be measured had turned yellow during lunch break. It was not difficult to deduce that this effect had probably been caused by sunshine accidentally falling on the Erlenmeyer flask with the solution and that the yellow material was nitrophenolate.

Excitement grew after we realized that the light-induced reaction occurs selectively with the *meta* compound and far less in the case of the *ortho* and *para* isomers.[155] This selectivity was in clear contradistinction to the classical rule of *ortho–para* activation. We decided to change our program completely and switched to the study of this novel phenomenon. It appeared to involve a heterolytic photoreaction, whereas at that time, photoinduced processes were normally associated with homolytic reactions and free-radical chemistry. Furthermore, the photoreaction showed *meta* activation, an unheard-of phenomenon that was contrary to the *ortho* and *para* interactions normally encountered in aromatic chemistry. This was the first observation of the remarkable oppositeness found when photoinduced processes are compared with similar thermal reactions.

Thus in the mid-1950s at Leiden we entered the field of photochemistry, in which organic chemists had a lot to learn. Our experiences with photoinduced nucleophilic substitution gave a direct warning that the concepts and rules of thermal reactions should not be used for photochemical processes without careful consideration. A similar sign soon came through the comparison of the photoinduced and thermal cyclization reactions in the vitamin D field.[107,110,111]

Photochemistry gradually became attractive to organic chemists, who contributed to widening the field by their keen interest in the detailed structure and reactions of medium-size and larger molecules, including those of biological origin. Moreover, there was the perspective of synthetic applications, as clearly shown in the industrial production of vitamin D. The highly esteemed photochemist W. A. Noyes, Jr., at his retirement from the chair of physical chemistry at Rochester in 1963, remarked with a good sense of humor (his father was a well-known organic chemist): "The strongly forbidden transition has occurred; organic chemists have overcome the barrier to photochemistry, traditionally the domain of physicists and physical chemists."

That the novel photoinduced process was not simple from a mechanistic point of view became clear through the first studies.[111,156,157] When ^{18}O was used as a label, the photoreaction of m-nitrophenyl phosphate with water was found to effect hydrolysis of the phosphorus–oxygen bond. With m-nitrophenyl sulfate, the difference in rates of photoreaction when compared to those of the *ortho* and *para* compounds is even more extreme; up to pH 14, only a reaction with water was observed. m-Nitrophenyl phosphate in solutions of pH higher than 12 shows an additional reaction with hydroxyl ions that leads to the splitting of the carbon–oxygen bond. Evidently the latter reaction constitutes a genuine aromatic substitution induced by activation of the aromatic system upon π–π^* light absorption (Scheme VI).

It remains a pity that lack of time did not allow a closer study of the first type of reaction: attack of the nucleophile at the phosphorus or sulfur atom. The reaction not only constitutes a theoretically interesting process; it also may prove valuable in applications as a means to photochemical phosphorylation (sulfurylation) of hydroxyl and perhaps other nucleophilic groups.

We chose to concentrate on the photoinduced nucleophilic aromatic substitution proper (S_NAr*). It proved to be a rather general reaction that occurs in the case of polycyclic and heterocyclic compounds and with a broad range of substituting and leaving groups.[158] Water, (polar) organic solvents, and liquid ammonia can be

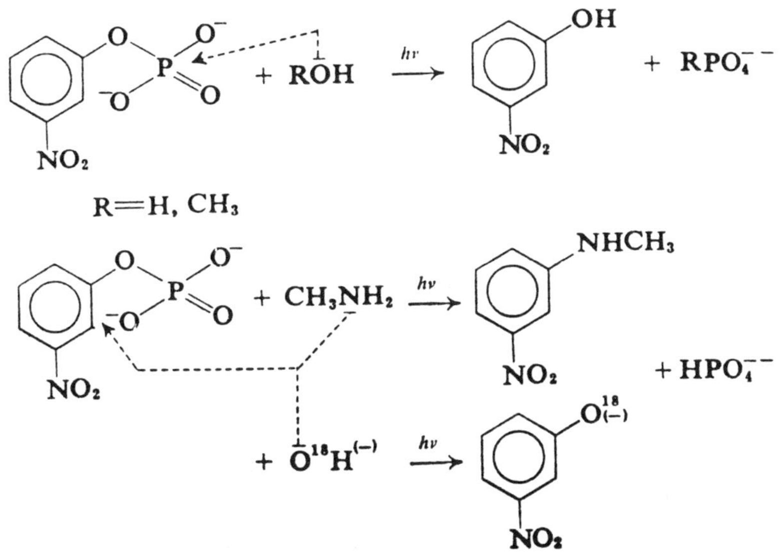

Scheme VI. *Nucleophilic photosubstitutions of* m-*nitrophenyl phosphate.*

used as media.[159] Hundreds of reactions have been described by authors from Leiden and other laboratories.[160]

The various aspects of the S_NAr^* reaction provided research subjects for the theses of some 20 students over a 30-year period. The research brought with it an instructionally attractive combination of the synthesis of the required compounds, photochemical experiments, kinetic studies, and theoretical interpretations combined with calculations of charge distributions and energies.

R. O. de Jongh certainly deserves special mention. After making the fundamental discovery in 1954, he constructed a reliable basis for further exploration by painstaking and critical studies in the new field before he felt justified to write his thesis.[161] In addition, I would like to pay tribute explicitly to former co-workers, later colleagues, who have been inspiring leaders of the photochemistry group at Leiden: the late M. E. Kronenberg, J. Cornelisse, J. Lugtenburg, and G. Lodder.

As a rule, nucleophilic aromatic photosubstitutions are clean reactions resulting in a good yield of product. Peculiar cases excepted, the quantum yield is largely independent of the wavelength of irradiation and the temperature. Although a few examples have been found in which the excited singlet starts the reaction with the nucleophile, mostly it is the π–π^* triplet of the aromatic compound that undergoes the substitution.

A simple scheme to serve as a first orientation is indicated in Figure 11. The aromatic compound is brought into an excited singlet state

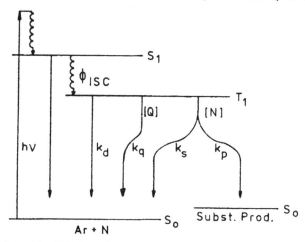

Figure 11. Simplified scheme for bimolecular nucleophilic aromatic photosubstitution via the triplet state ($S_N 2Ar^{3*}$).

by absorption of light. Relaxation to the thermally equilibrated lowest excited singlet is fast, and one of several pathways of deactivation is followed from there. One of these pathways leads via interconversion to the lowest $\pi-\pi^*$ triplet and reaction with the nucleophile to the formation of the ground-state product. Even in such a simple case, the efficiency of product formation yields direct information only on the ratio of its rate versus the rates of other deactivation processes determining the lifetime of the reacting excited species.

The entrance hall of Gorlaeus Laboratories, June 1983. The author with three former students: J. Cornelisse, H. J. C. Jacobs, and G. Lodder.

To know whether in the case of the *meta* compound (in comparison with the *ortho* and the *para* isomers) there is a real activation of
the substitution reaction proper, one has to eliminate the possibility that
a difference of the excited-state lifetime causes the difference of efficiency between the isomer photoreactions. A decision could be reached
by studying the reactions of compounds like 4-nitroveratrole (1,2-
dimethoxy-4-nitrobenzene), for which the efficiencies of substitution can
be compared at different positions of one and the same excited
molecule. Scheme VII shows early examples in which the preferred *meta*
activation is established as a contrast to the *ortho–para* activation of
thermal reactions.[157,162] The example of 4-nitroveratrole has an additional feature, in that the product of the one-step photoreaction with

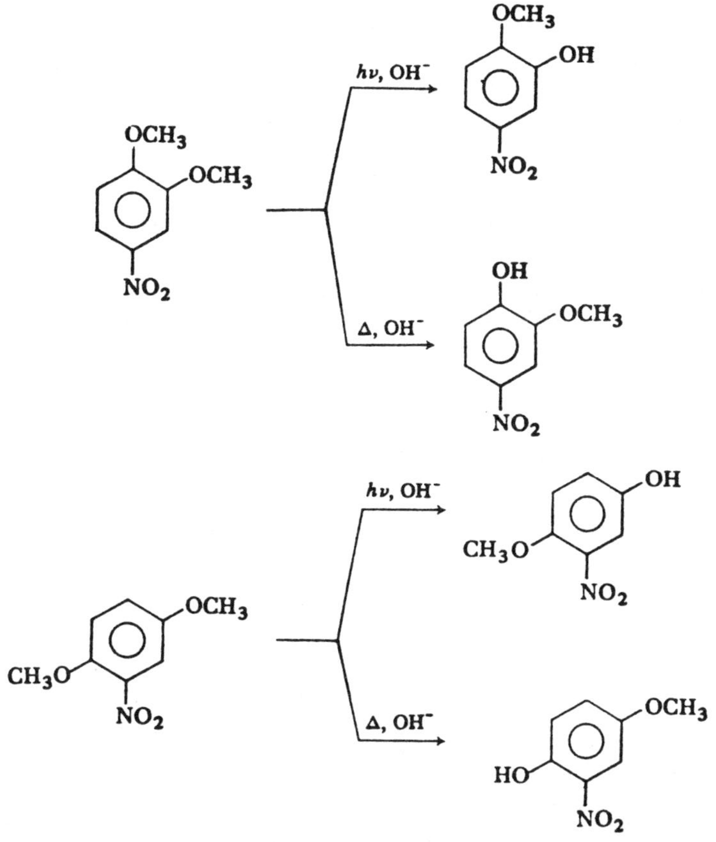

Scheme VII.   meta *activation in photoinduced nucleophilic aromatic
substitution versus* ortho–para *activation in thermal reactions.*

methylamine needs a much more complicated synthesis by thermal processes.[163]

In a later stage of the studies on the S_NAr^* reaction, the separate rate constants of bimolecular reaction with nucleophiles and the lifetimes of a few representative methoxy- and fluoronitronaphthalenes were determined (J. G. Lammers[164,165]). The rates of reaction of the methoxy compounds in the excited state are greater than those of the fluorides. However, at the same nucleophile concentrations, the nitrofluoronaphthalenes show a greater quantum yield of substitution because of overcompensation of the lower reactivities by a considerably longer lifetime of their triplets (Table IV).

Clearly, to reach sensible conclusions on reactivities of excited species in a photochemical conversion, we must break down the quantum yields into intersystem crossing efficiencies, lifetimes of the reacting species, and their specific rates of reaction. The results presented in Table IV and in Figure 11 constitute an example of the information that can be obtained in the study of photoreactions by scrutinizing via orthodox kinetic measurements and luminescence studies.

The reactions under consideration appear to represent authentic nucleophilic substitutions starting from the $\pi-\pi^*$ triplet state. Interestingly, 1-fluoro-6-nitronaphthalene in the triplet state shows reaction parameters that qualitatively and quantitatively are similar to those of

Table IV. Rate Constants of Nucleophilic Photosubstitutions of Some Fluoro- and Methoxynitronaphthalenes*

Compound		ϕ_{isc}	k_p	k_s	$\tau = k_d^{-1}$
	$+ OH^-$	0.8	9.6×10^7	8.8×10^7	2.3×10^{-7}
	$+ CH_3NH_2$	0.8	1.2×10^8	2.3×10^8	2.3×10^{-7}
	$+ OH^-$	0.8	1.0×10^8	1.5×10^8	2.5×10^{-7}
	$+ OH^-$	0.2	7.1×10^8	7.1×10^8	3.8×10^{-9}
	$+ CH_3NH_2$	0.2	1.2×10^9	1.2×10^9	2.5×10^{-9}

See Figure 11, page 48.

1-fluoro-3-nitronaphthalene. It seems as though, with $\pi-\pi^*$-excited polyaromatic compounds, the *meta* activation extends into the neighboring ring far more than substituent effects do in the ground state. We have encountered several examples of such extended *meta* activation in photoinduced aromatic substitutions.[166] With 7-nitro-2-fluoronaphthalene, triplet quenchers were found to have no influence, and an excited singlet-state reaction seems to occur.

A source of some mechanistic concern—more than compensated by synthetic fun—arose when the scope of the reaction was extended and nucleophilic photosubstitutions were effected in liquid ammonia (A. van Vliet[167–169]). The now-familiar pattern of *meta* activation was found with nitroaromatic ethers. With other aromatic compounds, *ortho–para* orientation by the nitro group was observed (Scheme VIII). The occurrence of an *ortho–para* pattern, after we had become accustomed for some years to seeing *meta* activation in photosubstitution, reminded us that one has to adapt rules and principles again and again as a result of the novel phenomena that the experimental chemist encounters today as often as the alchemists did in their time. As a working hypothesis in this case, it was considered that with the high concentration of the reactant in liquid ammonia, reaction selectivity in the excited state might be decreased and product formation could be oriented mainly in the subsequent "dark" part of the process. Merging resonance stabilization during the formation of the ground-state product could become a directive factor in favor of *o*- and *p*-nitroanilines.[170]

Still greater unorthodoxy was to be uncovered shortly. Gradually the suspicion grew that methoxy and other electron-donating groups might exert an activating influence on nucleophilic (!) photoinduced

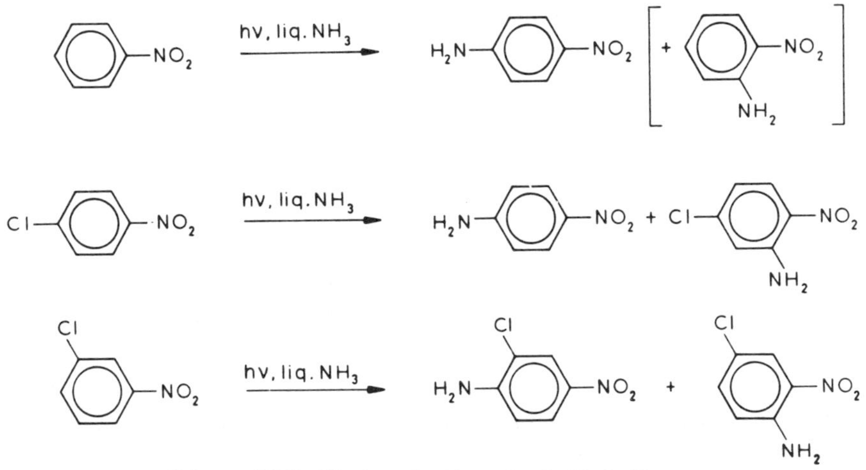

Scheme VIII. Photoamination of substituted benzenes.

substitution. An early indication was with *p*-nitroanisole; in the reaction with hydroxyl ion and with amines, substitution of the nitro group was found to be the major process.[171–173] In 2-halogeno-4-nitroanisoles, the halogen atom exhibits a remarkable reactivity with respect to nucleophilic photosubstitution.[174,175] The *ortho–para* activation by electron-donating groups was confirmed through a special investigation of 1,3,5-trimethoxybenzene by Lok,[176] who established in that compound a smooth substitution of hydrogen by nucleophiles (Scheme IX). El'tsov and his colleagues[177] reported the efficient substitution of the halogen in *p*-halogenoanilines. *ortho–para* activation in the case of anisole was found by Nilsson,[178,179] a co-worker of Eberson at Lund, who investigated the photoreactions in connection with the study of electrochemical processes.

I have to admit that in certain phases of the research I was captivated by the unexpected elegant reactions and remained engaged in collecting examples of novel photoinduced processes, rather than spend much effort in early rationalization. Certainly, valuable orientation rules could be formulated: *meta* activation by electron-attracting groups; *ortho–para* orientation by electron donors; merging resonance stabilization as dealt with previously; and a fourth regularity, α reactivity (preference for the α position in naphthalenes and other condensed polycyclic aromatic compounds).[170,180,181] Even so, there remained a feeling of uncertainty on account of the conspicuous diversity of reaction courses, depending on the system under consideration.

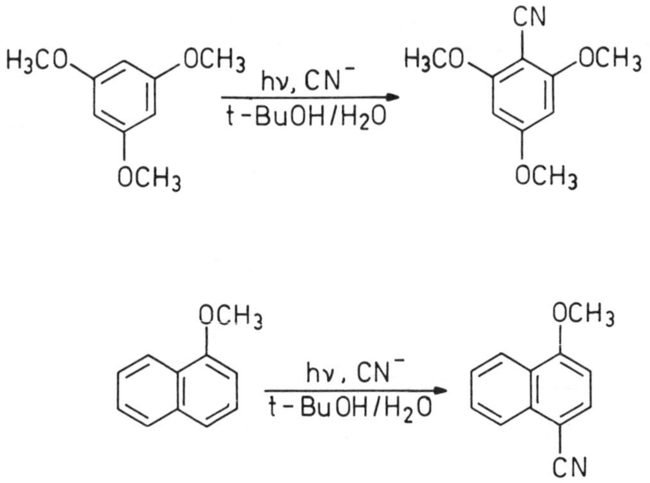

Scheme IX. ortho–para activation by methoxy substituents in nucleophilic (!) aromatic photosubstitution.

In the mid-1970s, the complexities of the field could be unraveled by the realization that several mechanistically different categories of S_NAr^* reactions are possible. This thought process was catalyzed during the preparation of the text of a lecture for the 1975 IUPAC (International Union of Pure and Applied Chemistry) congress in Jerusalem. As I was told later, I had somewhat misunderstood the intention of the lecture. The talk was to have been a presentation of vitamin D photochemistry, rather than aromatic photosubstitution. However this may have been, the request inspired J. Cornelisse and me to reflect on the whole of the experimental data. We recognized several distinct categories of aromatic photosubstitution and proposed a classification based on mechanistic aspects (Scheme X).[170] The consistent totality gave us confidence that we were on the right track to an understanding of diverse aromatic photosubstitutions within a mechanistic framework.

The system adopted for the classification has a rational basis. The excited molecule intrinsically has high internal energy, and it seems reasonable to attribute a great reactivity to it. This reactivity would include the tendency to undergo authentic bimolecular substitution to form a ground-state product (S_N2Ar^*). Then the $\pi-\pi^*$-excited molecule boasts a promoted, loosely bound electron and an electron hole. We therefore should incorporate the possibilities that the loss of an electron might yield a radical cation ($S_{R+N}1Ar^*$) and the gain of an electron could mean formation of a radical anion ($S_{R-N}1Ar^*$). Moreover, the classification system should include the pathway via dissociation into ions (S_N1Ar^*)[182] and intramolecular charge separation (S_NICSAr^*). The latter mode of reaction was established by the elegant investigations on biphenyl derivatives and related compounds by O. B. Shadid.[183]

A gratifying result of this classification based on different reaction pathways was that such strange anomalies as promotion of nucleophilic substitution by electron-donating groups no longer cause us uneasy feelings. The initial photochemical step in that case is the transfer of an electron to the solvent or to a ground-state molecule.[184,185] The ensuing substitution consists of a thermal process known in radical cation chemistry. Nilsson at Lund[178] and J. den Heijer at Leiden[186] have shown that the product composition of the photoinduced and the electrochemical radical cation processes are convincingly similar.

The processes occurring via the primary formation of radical anions, first investigated by Kornblum[187] and Russell,[188] were introduced in the aromatic field by J. F. Bunnett and his co-workers.[189,190] The various reactions of this class are theoretically interesting and show a great synthetic potential. A few possible examples have been found for the S_N1Ar^* type of substitution besides the case tentatively indicated in Scheme X.[191,192] Some of the light-promoted reactions of diazonium salts may belong to this category.

And then of course we come back to the origin of our investigations: the true bimolecular substitution of the photoexcited aromatic molecule (mostly in the triplet state, $S_N 2Ar^{3*}$). Two lines of interpretation and further investigation on this "old faithful" may be

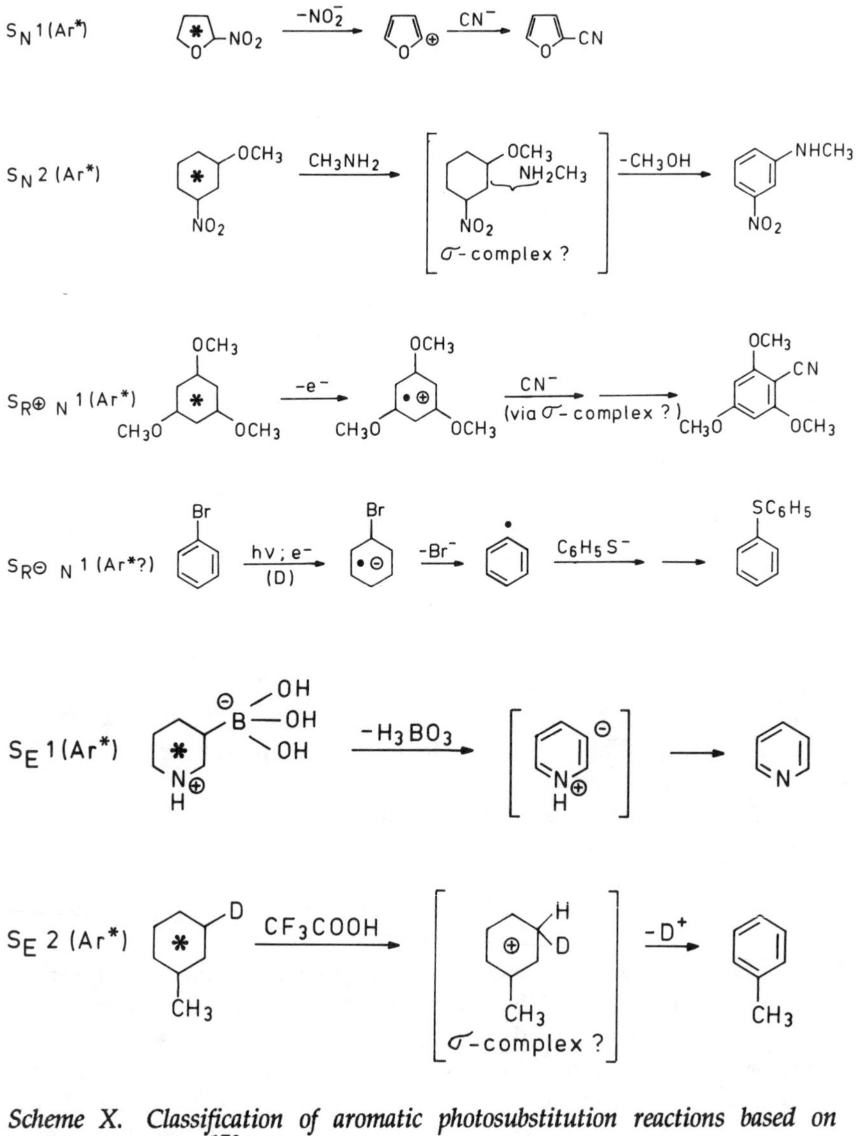

$S_N 1 (Ar^*)$

$S_N 2 (Ar^*)$

σ-complex?

$S_{R\oplus N} 1 (Ar^*)$

(via σ-complex?)

$S_{R\ominus N} 1 (Ar^*?)$

$S_E 1 (Ar^*)$

$S_E 2 (Ar^*)$

σ-complex?

Scheme X. Classification of aromatic photosubstitution reactions based on mechanistic criteria[170].

summarized at the end of this section in order to give an indication of future developments.

The first line of approach is to use the impressive recent developments in photochemical and spectroscopic methodology. By illuminating the system with strong laser flashes, we can bring a good number of molecules into an excited state in a very short time. Their fate is then followed by various spectroscopic methods. In principle, and often in practice, the lifetime and nature of short-lived intermediates can be determined in this way. Pathways of deactivation, including product formation, can be traced and compared with the results of the calculation of energies and properties of the species concerned.

At Leiden, investigations along this line were initiated at an early stage by Cornelisse and De Gunst.[193–195] As a subsequent study of the S_N2Ar^* reactions, Tamminga, in collaboration with C. A. G. O. Varma's spectroscopy group, investigated the photohydrolysis of 3,5-dinitroanisole.[196,197] This clean bimolecular process gives 3,5-dinitrophenol in good chemical yield with a quantum efficiency of 0.4 in 0.01 M NaOH.

Activation by a nanosecond UV flash produces intermediates of different lifetimes (Figure 12). First an absorption appears with a maximum at 435 nm, which is caused by a species with a lifetime of the order of 10–100 ns (depending on the concentration of hydroxyl ions) that can be identified as the triplet. Evidently this triplet species reacts with the nucleophile to form an excited complex with absorption maxima at 395, 570, and 650 nm and a lifetime of the order of microseconds. This complex, or mixture of complexes, then transforms into several ground-state σ complexes. These generally dissociate to form the starting material again. However, the carbon-1 σ complex with the methoxy group and the nucleophile at the tetrahedral ring atom has a substantial probability of forming the substitution product, 3,5-dinitrophenolate.

The simplified description given here (Figure 12) is almost unfair, because it does no justice to the extensive excellent research and the detailed picture achieved. However, one crucial result should stand out from the presentation: Tamminga succeeded in demonstrating visibly the existence of σ complexes as intermediates. These complexes were postulated almost from the beginning of the research but never convincingly observed in earlier investigations.

Tamminga's study exemplifies the exploration of the reaction pathways of aromatic photosubstitution with attention focused on the intermediates and their interconversions. Of course the organic chemist also likes to rationalize the characteristic regiospecificities observed with the S_NAr^* processes. For some classes, the selectivities are those of the radical cations or anions that are formed from the photoexcited species and that react via a series of dark reaction steps to form substitution

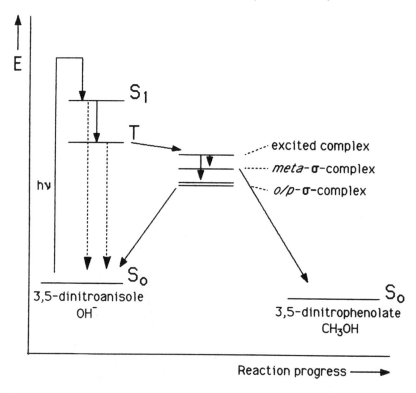

Figure 12. Intermediates in S_N2Ar^* of 3,5-dinitroanisole, traced by means of the nanosecond-flash technique.

product. Here the time-honored approach and the experiences of ground-state chemistry can be used as a basis.

The crucial question remained whether in the true S_N2Ar^* process the selectivity is also effected in the dark reaction (formation of the product from the σ complexes) or during the first phase of the processes, the relaxation of the system in and from the excited state. Van Riel and Ferrier, under the guidance of Lodder,[198–201] made a closer investigation of the question. Selectivities based on the dark reactions starting from the σ complexes were eliminated by making the substituting and the leaving group (OCH_3^- in methanol) chemically equal.

The reactions of a series of nitromethoxyaromatic compounds were followed by using a ^{14}C label. These symmetric methoxy substitutions with nitroanisoles revealed a clear *meta* orientation, although it was somewhat less pronounced than that observed with substitution by hydroxyl ion. Therefore, although part of the regioselectivity may be achieved during the dark conversions of σ complexes, it seems proven that the *meta* activation largely originates with the conversion of the excited species into the ground-state σ complexes.

More specifically, we can assume that the *meta* orientation may originate with the excited complex that is formed when an aromatic compound in the excited state encounters a nucleophile and, more probably, with the subsequent deactivation of the (mixture of) excited complexes to the ground-state σ complexes. Several factors in the deactivation favor the transition to the σ complex that in the ground state leads to the *meta*-substituted product. The ground-state σ complex leading to the *meta* isomer has a higher energy content than those corresponding to *ortho–para* substitution. Calculations suggest that the reverse relation holds with excited-state complexes of comparable geometry (*see also* P. Wagner[202]).

Evidently the energy gap for deactivation to the ground-state energy hypersurface is considerably smaller for the route that leads to the *meta* σ complex than for the *ortho–para* substitution (Figure 13). The probability of crossing the energy gap decreases exponentially with the increasing magnitude of the gap. This situation may represent an

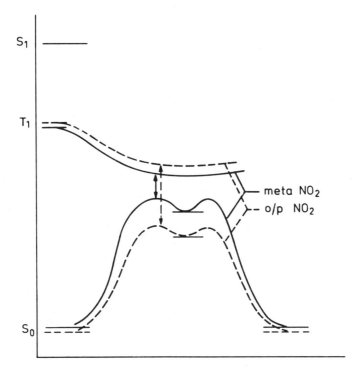

Figure 13. S_N2Ar^* reaction; schematic two-dimensional representation of pathways along multidimensional surfaces; meta activation.

important factor that directs the reaction course to *meta* substitution, which is not favored in ground-state chemistry. Certainly this description is greatly simplified. However, it gives the satisfactory feeling that it is related to the rationalization used with the other early example of opposite specificities of ground and excited states — the cyclization of hexatrienes in vitamin D chemistry.

However, there is more to this description. One of the intriguing features of the S_N2Ar^* mechanism is the observation that the nitro group has a much stronger effect than the cyano group. This difference parallels the lowering of the energy of the triplet (E_T) of aromatic compounds by various substituents (E_T benzene, 85 kcal/mol; E_T nitrobenzene, 60 kcal/mol; and E_T benzonitrile, 80 kcal/mol). The acetyl group is another electron-attracting group that substantially decreases the deactivation energy gap. Indeed, a smooth *meta*-oriented photosubstitution is observed with the acetoanisoles.

Only a few examples of electrophilic aromatic photosubstitution have been found (Scheme X). This contrasts to normal aromatic chemistry, for which electrophilic substitution is one of the best known processes. Evidently the energy maxima of the trajectories to electrophilic substitution are low. This situation may contribute to making the hypersurface of reaction too far down to be reached from the excited state. Certainly other factors play a role. Many electrophiles are efficient excited-state quenchers. Moreover, it is not easy to find appropriate media with sufficient strong-electrophile concentration to create a good chance for the short-lived excited aromatic compound to encounter an electrophilic reaction partner.

These ideas represent lines of overall reasoning. In the future, they should be complemented or even substituted by more solid approaches to investigation of energy hypersurfaces. Exploration of excited-state and ground-state energy hypersurfaces and their interactions is essential for a satisfactory description of photoinduced reactions and thermal processes. We may expect the study of excited-state processes to develop within the next 10–20 years to the extent that quantitative predictions will become possible.

The study of photoreactions is attractive beyond its potential to lead to sophisticated synthetic and other applications. Great value has to be assigned to the possibility that many systems can be brought by photoflash excitation into energy levels that practically can not be reached via classical thermal procedures. A telling example is presented in the work of Tamminga, [196,197] in which σ complexes (essential but not perceptible in thermal aromatic chemistry) could be generated in such concentrations that direct observation became possible.

This mutually supportive relationship between photochemistry and ground-state chemistry is indicated by an extension of the picture of Klaus Müller in his illuminating article in *Angewandte Chemie*.[203] The picture, which illustrates one of the essential problems in the study of ground-state processes, is comparable with Figure 14. The present version includes the exploratory parachuting down that symbolizes the help that may be had from photoinduced processes in theoretical and experimental organic chemistry.

This section, dealing with some aspects of the photochemical research at Leiden, aromatic photosubstitution in particular, was written essentially from memory. Many valuable contributions of workers in the Leiden photochemistry group have not been done justice. This is inevitably true also for the investigations of colleagues working in the fields under consideration at other centers. We enjoyed the privilege of direct cooperation with a few of them: E. J. Poziomek (Aberdeen), C. Párkányi (El Paso), G. Fráter (Zürich), and G. G. Wubbels (Grinnell), who stayed as visitors at Leiden for a longer period.

I hope that the sieve of restricted memory has highlighted essential aspects and innovations and that the stories presented will create an unembellished and useful impression of the best parts of this wonderful world of photochemistry.

Photochemists all over the world feel that they belong to a large family of good relatives. This feeling grows for instance, from effective

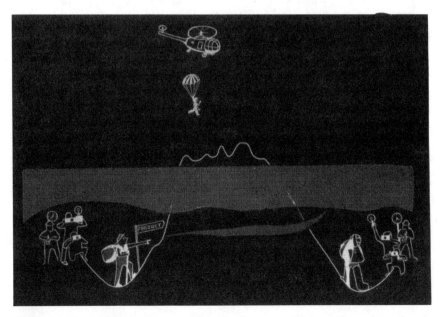

Figure 14. Approach from a high level; new vistas and support in the exploration of transition pathways.

Probably taken at the Gordon Conference, 1979. N. J. Turro is behind Havinga.

mutual support and the exchange of results and ideas at the various conferences. It is indicated also by the viability of organizations of photochemists in the Old World (EPA), the New World (IAPS) and last but certainly not least in the East (JPA).

I am grateful for having found and regularly encountered many of my best friends in the communities of photochemists, including the ATP (Association of Tennis-Playing Photochemists). I hope that photochemists and photochemistry will continue to cultivate the pleasant atmosphere of mutual appreciation and cooperation that I have enjoyed so much during the past 35 years.

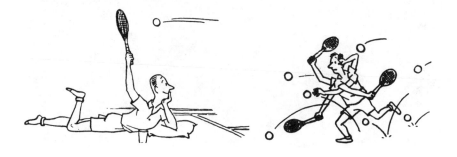

Reactions in the ground state—and the excited state.

With W. C. Agosta and J. Kossanyi. Probably taken at the Leuven Photochemistry Conference, 1979.

Shorter Lines of Research

During the first decade following the reopening of Leiden University in 1945, there was a great influx of students who had not been able to start or to continue their studies because of the war and occupation. These students generally were strongly motivated, passed their examinations within a short time, and often continued doing research for some 3 to 5 years to earn their doctorates. The relatively large group of those who chose to specialize in organic chemistry naturally included a great variety of scientific background and ambition. Some were primarily interested in theoretical and mechanistic aspects. Others felt attracted toward synthetic challenges. Quite a few had been inspired by the marvelous chemistry observed in nature or by the applications of chemical knowledge and insight in medicine and in technology.

Organic research generally includes several of these aspects: synthesis of the compounds necessary for the study, examination of the conformational behavior and reactivities of the molecules by physical and chemical methods, theoretical discussion, and calculations. Even so, it appears particularly stimulating when the first real research theme is chosen on the basis of personal preference. A broad range of research fields, offered as a basis for the students to make their own choice, fosters a direct relationship and feeling of responsibility for the ensuing investigations.

In accordance with the traditional Netherlands system, there was just one professor and one reader ("lector") of organic chemistry at Leiden. Senior colleagues of the faculty anticipated that I, by education and personal interest, would feel sympathetic toward developing research in various directions. These diverse investigations could be expected to become stimulating and fruitful to each other sooner or later.

Concentration of research on a limited number of fields was achieved eventually in such a way that the students could still choose a subject according to personal preference. Four of these fields that have been elaborated during 30 or more years are dealt with in the other sections. This section presents an overview of two of the shorter lines of research. Each of these was developed by several co-workers until a (for that time) satisfactory state of knowledge had been obtained.

Styrenes and Stilbenes: Transmission of Substituent Effects as a Function of Molecular Nonplanarity

This research subject had its roots in the early investigations on plant growth hormones in the Utrecht organic chemistry laboratory. As a student under the guidance of Kögl's assistant, Veldstra, I prepared a specimen of indole-3-acetic acid that was required as the identity standard for the newly discovered plant growth hormone, heteroauxin. A range of compounds of similar structure and geometry subsequently were found to show hormonal activities like indoleacetic acid.

A suggestive result was observed with the isomeric E- and Z-cinnamic acids. The sterically crowded nonplanar molecules of the Z compound possess considerable plant growth activity. The E isomer, which has no substantial hindrance to adopting the resonance-favored planar form of the molecules, shows no activity at all. In the years during and just after the war, when I worked at the veterinary faculty at Utrecht, there was an effective cooperation with Veldstra, who had become research director of the Amsterdamse Kininefabriek. Veldstra had put forward the attractive hypothesis that one of the factors determining whether a carboxylic acid containing an aromatic moiety shows plant-growth-influencing activity might be related to a preferential nonplanarity of the molecule.[204,205]

Planar conformation of (E)-cinnamic acid; sterically hindered nonplanar Z isomer.

At Leiden, an extensive study of the geometries and the properties of Z- and E-styrene derivatives started with the thesis research[206–208] of R. J. F. Nivard, who later was to become the first professor of organic chemistry at the University of Nijmegen (1954–1984). The methods of investigation available at that time, applied by Nivard and subsequently by O. J. Mattray,[209] were UV absorption spectroscopy and determination of dipole moments and dissociation constants.

Nivard's results and those of G. Riezebos[210,211] and W. H. Laarhoven[212–215] on stilbenes seemed to be consistent with the notion of the preferred planarity of the molecules of the E-compounds and the sterical hindrance to planarity of the Z isomers. But what remained doubtful was whether the transmission of polar effects exerted by substituents in the aromatic ring is stronger through the planar π-electron systems than through the sterically crowded nonplanar molecules. It was no wonder that this doubt caused uneasy feelings. We had expected to find positive indications comparable with (although smaller than) the classical examples of steric inhibition of resonance with, for instance, substituted p-nitroanilines and acetophenones.

It could be argued that the methods used in our studies were not optimal. Ultraviolet absorption is determined by the electronically excited states, as well as by the ground state. The interpretation of the values of the dipole moments for geometrically complicated molecules is far from easy. The dissociation constants are measured in polar media and may be influenced by "field" effects. Therefore it was decided to elaborate on the problem, the more so as accurate IR techniques became available.

Kronenberg (thesis 1962) extensively used IR spectroscopy and measured the carbonyl stretching frequencies of a series of substituted benzalacetones (benzylideneacetone) in carbon disulfide.[216–218] The conclusion again had to be that the transmission of substituent influences in the benzalacetones is hardly different for the (planar) E and the sterically hindered Z isomers. The values of the dipole moments indicate that the carbonyl group of the Z benzalacetones is indeed considerably out-of-plane, as is suggested by the consideration of molecular models.

All experimental results obtained thus far suggested that, against expectation, the transmission of polar substituent effects is practically the same for the sterically hindered nonplanar styrenes and the unhindered compounds. Therefore, an investigation was undertaken in which the full potential of chemical structure variation was applied. This finalizing of the studies was effected by M. de Vries.[219]

A considerable part of De Vries' meticulous synthetic work could not bear daylight because of the liability of the compounds to photochemical isomerization. The syntheses made available an extensive series of substituted (E)- and (Z)-cinnamic ethyl esters, as well as the phenylpropiolic esters (including the tricky p-amino compounds). For comparison, the ethyl 2,6-dimethyl-(E)-cinnamates (steric crowding), the phenyl acetates, and the dihydrocinnamates (separation by a methylene or ethylene group) were also measured. Detailed discussions of the IR and NMR data, as well as the syntheses and the interesting chemical reactivities of many of the compounds, are in the thesis.[219]

Two definite conclusions emerge from the carbonyl stretching frequencies (in carbon tetrachloride) with respect to the transmission of electronic substituent effects through sterically hindered nonplanar and unhindered styrenes (Table V):

Table V. Carbonyl Stretching Frequencies of Various
Carboxylic Acid Ethyl Esters (cm^{-1})

Compounds	p-NH$_2$	p-OCH$_3$	m-NH$_2$	H	m-NO$_2$	p-NO$_2$
Ethyl benzoate	1717	1722	1726	1728	1736	1735
Ethyl trans-cinnamate	1716	1719	1720	1722	1726	1728
Ethyl cis-cinnamate	1724	1727	1731	1732	1733	1733
Ethyl 2,6-dimethyl-trans-cinnamate	1718	–	–	–	–	1728
Ethyl phenylpropiolate	1713	1716	1718	1718	1723	1723
Ethyl phenylacetate	1744	–	–	1745	–	1747
Ethyl hydrocinnamate	1743	–	–	1744	–	1744

1. As appears from the measurements of the 4-substituted (E)- and (Z)-cinnamic esters and the corresponding 2,6-dimethyl-substituted E compounds, transmission of polar substituent effects is practically not influenced by hindrance of coplanarity.

2. The influence of mesomeric effects is much stronger with (E)- and (Z)-cinnamic esters and with phenylpropiolic esters than with phenylacetic and phenylpropionic esters.

These conclusions seemed to be experimentally well based. Further elaboration had to await the development of novel methods of approach. It is indicative of the rate of growth of our science that such methods appeared within two decades. Now in 1987 the problems can be dealt with effectively by making series of specifically labeled (^{13}C and ^{2}H) compounds and investigating them by the highly informative NMR methods. Again the combination of the synthesis of compounds that are isotopically labeled at certain positions and advanced instrumental analysis effects significant progress. In this way, the situation at each of the carbon atoms can be explored, and a detailed picture of the conformation and charge distribution can be obtained. Current advanced theoretical calculations also contribute to a satisfactory description.

The impressive potential of the new methods can be seen from the results recently obtained in the field of rhodopsin research[220,221] and with substituted hexatrienes.[154,222] In the resonance descripton of nonplanar conjugated systems, dipolar limiting structures of the type shown might make a somewhat more substantial contribution than formerly assumed. Possibly in the not too distant future the curious data obtained with some styrenes and stilbenes briefly indicated in this overview will inspire a penetrating study that will arrive at a detailed description of the conformational features and the electronic organization of such molecules with extensive conjugated systems.

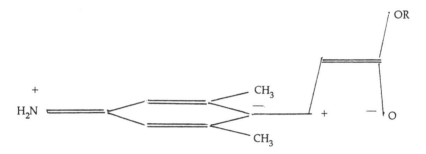

Polar resonance structure possibly contributing in the case of nonplanar conjugated system.

Aromatic Nitroso Compounds and Quinone Monoximes: Chameleonic Species

One of the other short lines of research pursued at Leiden during the years 1946–1960 had to do with the often-disputed behavior of nitroso compounds, nitrosophenols in particular. Controversial isomerization phenomena had been reported in the literature for several representatives of this class of compounds. Often the problems had been tackled either from the "chemical" point of view (synthesis) or by "physical" measurements, and not so much by integration of the two methods of approach.

We embarked on a closer study of the nitrosophenols in 1946 because we felt that the time was ripe for investigating these compounds by the rapidly developing methodology of physical organic chemistry.[223] Aromatic nitroso compounds show characteristic absorption in the ultraviolet (visible) and the infrared wavelengths, and they possess substantial dipole moments. As a rule, they can be obtained in crystalline form and be analyzed by X-ray diffraction. The physical methods of investigation should be complemented by a systematic variation of molecular structure. With the aromatic nitroso compounds this can be realized through rather delicate but not prohibitively complicated syntheses. The field thus seemed to offer excellent possibilities to young researchers for training themselves in various aspects of organic chemistry and for collecting results that could earn them the doctorate.

Tautomerism of Nitrosophenols–Quinone Monoximes. Our endeavor in this field was given a good start by one of my first co-workers at Leiden, A. Schors, later to become scientific director of the central research laboratory of the Netherlands Organization of Applied Research (TNO). He possessed the necessary skill and ambition in organic synthesis, combined with a good physical and theoretical background. A thorough study was made of a series of quinone monoximes–*p*-nitrosophenols and the phenomena connected with their tautomerization.[224–226] The ensuing results, which today may well constitute just a subject in an introductory organic chemistry course, eliminated some misleading elements from the earlier descriptions and established the following basic principles:

1. In polar solvents the two tautomeric species (nitroso and oxime forms) are in a dynamic equilibrium via their common (mesomeric) ions (Scheme XI).

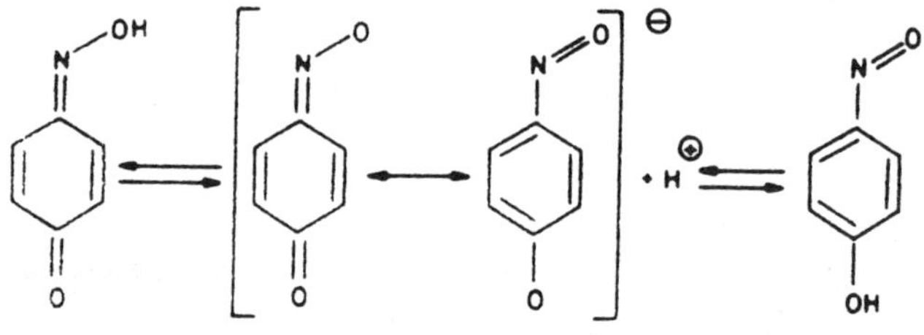

Scheme XI. Equilibration of nitrosophenol–quinone monoxime via the common (mesomeric) ion.

2. The quinone oxime form mostly predominates at equilibrium in solution. The crystalline phase generally consists of intermolecularly hydrogen-bonded quinone oximes.

A remaining anomaly was clarified by Schors and his successor in the field, A. Kraaijeveld.[227–229] Two products of nitrosation of 3-chloro(bromo)phenol are reported in the literature by several authors.[230,231] There is a stable high-melting compound (mp 184 °C), generally described as the 3-chloroquinone-4-oxime, and a controversial light-sensitive product (mp 142 °C) that some authors consider to be the 3-chloro-4-nitrosophenol isomer.

Schors had no great difficulty in confirming that the high-melting compound is pure 3-chloroquinone-4-oxime. However, the low-melting product appeared to be a 1:1 molecular complex. One component is 3-chloroquinone-4-oxime. The other component, yellow and light-sensitive, could be isolated in a pure form. Straightforward synthesis proved it to be 3-chlorobenzene-6-diazooxide (mp 124 °C).[227–229]

3-Chlorobenzene-6-diazooxide.

As a check, the enigmatic product (mp 142 °C) was prepared by crystallizing a 1:1 mixture of the oxime and the diazooxide. Elegant confirmation of the conclusions came from an infrared study by Philbrook and Getten.[232] The light sensitivity of the diazooxide (which shows absorption around 400 nm) is related through a sequence of conversions (Süs,[233] De Jonge[234–236]) with an initial light-induced elimination of nitrogen. The nitrosation products of 3-bromophenol were found to be analogous to those of the chloro compound.

Thus a long-standing controversy had been clarified. Moreover, at Leiden we had experienced our first encounter with a photochemical process. It served as a prelude to a major development in vitamin D chemistry and aromatic photosubstitution.

Dimerization. Besides tautomerization to oximes, the nitroso compounds show a capacity for dimerization (Scheme XII). Analysis by X-ray diffraction[237,238] proved the dimers to have the structure indicated in Scheme XII, as originally proposed by Hammick et al.[239] Infrared studies and determination of dipole moments[240–243] suggest that the dimers may have a centrosymmetric *anti* (*E*) structure or a *syn* (*Z*) configuration. The *anti* N_2O_2 moiety is planar. The monomeric nitroso derivatives are characteristically greenish-blue because of the $n–\pi^*$ absorption. This absorption (molar extinction ϵ generally around 45^{244}) enables one to estimate the percentage of monomer present in a solution.

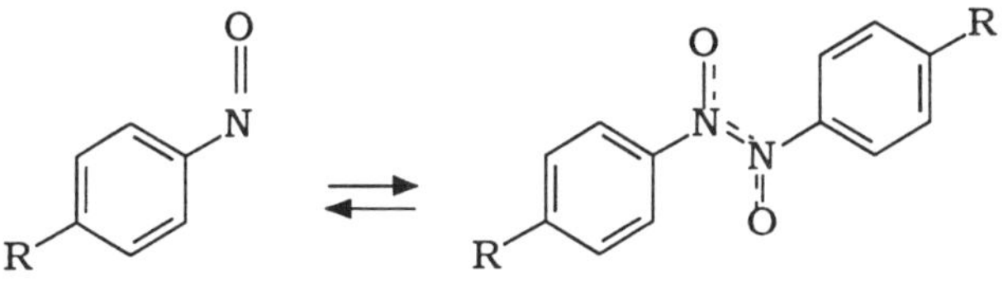

Scheme XII. Dimerization of nitrosobenzenes.

Gratifying results were obtained by investigation into the influence of substituents in the aromatic ring of nitrosobenzenes (Mijs,[243,246] Hoekstra[245]). It was not difficult to foresee and establish that electron-donating groups, at the *para* position in particular, favor the monomeric form by resonance interaction. Compounds like 4-methoxy-, 4-methyl-, and 4-iodonitrosobenzene occur as monomers in solution, as well as in the crystalline phase.

The prediction that *ortho* substituents would promote dimerization by their steric influence seemed somewhat paradoxical.[247] Steric crowding should reduce the feasibility that the monomer would assume the resonance-stabilized planar conformation. Aromatic rings of the dimers are not coplanar with the N_2O_2 moiety. Extensive investigations with several series of compounds established[243,245,246] that substituents at the *ortho* positions promote the occurrence of the dimeric species through their steric effects. For instance, 4-iodonitrosobenzene in chloroform solution is practically 100% monomeric (ϵ 45). However, simple substitution of two methyl groups at the *ortho* position effects a concentration-dependent dimerization (ϵ of 2,6-dimethyl-4-iodonitrosobenzene in chloroform: 17 at 0.01 mol/L; 26 at 0.002 mol/L). The nice interplay of steric effects and of stabilization through resonance could be observed with a great number of aromatic nitroso compounds.

Valuable information often was obtained through X-ray diffraction analysis. The organic chemistry department considered it essential to have continuous cooperation with an expert crystallographer. C. Romers was deeply interested and took part in the investigations of some of the organic groups. Moreover, he gave personal guidance to organic chemists as they completed the X-ray analysis of the compounds they were studying. I am very grateful that this pleasant and productive cooperation could be realized at Leiden. Such a direct relationship between organic chemistry and diffraction analysis should be a normal feature at a university laboratory to complement the invaluable contributions of molecular spectroscopy.

At the end of this short account of some of our investigations dealing with aromatic nitroso compounds, I should mention explicitly the extensive studies of other colleagues. B. G. Gowenlock, W. Lüttke, and many others made important contributions to the colorful chemistry of nitroso compounds.

E–Z Isomerization. In addition to tautomerism and association, the nitrosophenols–quinone monoximes can play yet another game. This is the formation of E–Z (*anti–syn*) isomers. We encountered several examples of isomerism that had been ascribed to oxime–nitroso tautomerism, but then appeared to be E–Z isomerism (Umans, Fischmann).[248,249] Here again, the combination of spectral data and X-ray analysis usually gave a definite answer. I will describe just one interesting case in which we were left with an intriguing and only partly definite answer.

2-Chloro(bromo)-5-methylbenzoquinone.

Some 90 years ago F. Kehrman described two isomers that he had obtained by oximation of 2-chloro(bromo)-5-methylbenzoquinone.[250] Controversy about this observation arose later. At Leiden, following Kehrman's oximation procedure, we experienced no difficulty in isolating a nicely crystallizing α form and a fibrous β modification, obtainable only as very small crystals. Fischmann, whose organic chemistry study had included crystallographic training by C. H. MacGillavry at the University of Amsterdam, made extensive X-ray diffraction studies under the guidance of Romers. These investigations proved the α form to be the *anti* isomer of 2-chloro-5-methylbenzoquinone-4-oxime.[249,251,252] The crystal contains long sequences in which the quinonoid molecules are lined up and intermolecularly hydrogen-bonded, as indicated in Figure 15.

The β form shows significant similarity to the α form in its UV and IR spectra. A decisive X-ray analysis has not been made because appropriate crystals could not be obtained. However, combination of the chemical reactions with data from fiber and powder diagrams, from UV and IR spectroscopy, and from molecular model building leads to the *syn*-2-chloro-5-methylbenzoquinone structure. In the crystal, the *syn* isomers are supposed to build helices with six hydrogen-bonded units per turn, as indicated in Figure 16. The helices on the inside are clad by methyl groups.

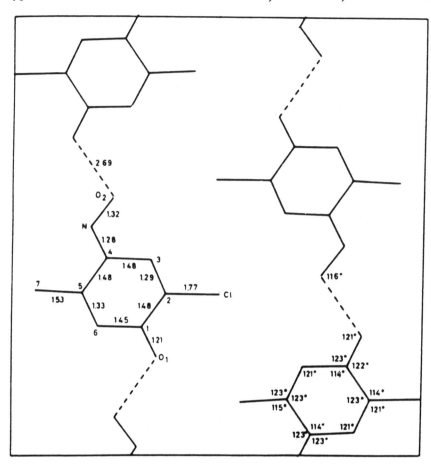

Figure 15. Packing of molecules of α-2-chloro-5-methylquinone-4-oxime in the crystal.

Even today an organic chemist may understand our gratification 30 years ago when one of the oldest color reactions in chemistry (the characteristic blue staining by a solution of iodine) was applied as a check on the model. Fischmann[249] decided to try this reaction that is so clearly positive for amylose, which is known to consist of molecular tubules possessing a hydrocarbon lining. He observed a beautiful blue color when he combined on paper and dried a solution of the β form of 2-chloro(bromo)-5-methylbenzoquinone-4-oxime and a solution of iodine in alcohol. The α isomer does not show this color reaction.

This was one of the simplest and most enjoyable chemical experiments I can remember. However, in science there is no definite happy end to any story. It still remains hard to swallow the fact that, unlike

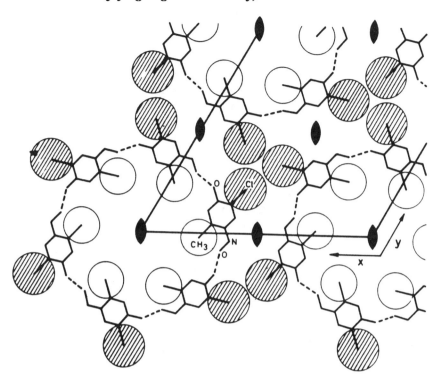

Figure 16.   β-2-Chloro-5-methylquinone-4-oxime; probable arrangement of molecules in the crystal.

our experience with other E–Z quinone monoximes, we did not observe interconversion of the α and the β forms in solution. Kehrman reports such a conversion, but it has not been observed by other investigators. We may not have chosen the optimum conditions, but it seems beyond doubt that the equilibration, to be expected on the basis of the equation of Scheme XIII, does not occur easily. Relevant factors are the low per-

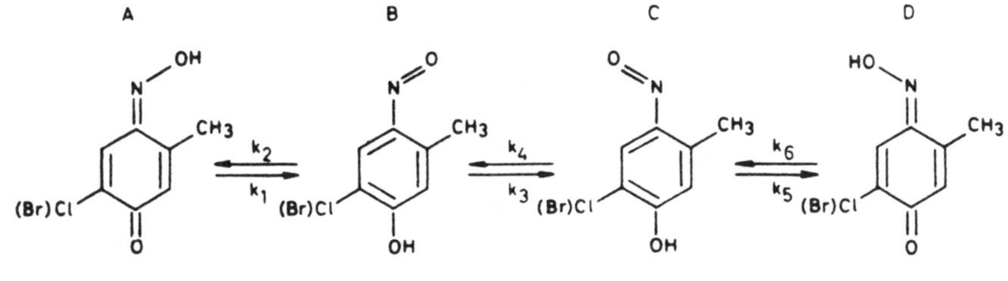

Scheme XIII.   Possible cis–trans isomerization of 2-chloro(bromo)-5-methylquinone-4-oxime.

centage of the nitroso protomer in the equilibrium (<1% by spectroscopic evidence) and the barrier against rotation around the C–N bond
that has substantial double-bond character.[253] Still, I feel there remained
an anomaly to be solved by more detailed investigation when we closed
the Leiden research on nitroso compounds around 1960. This field of
investigation had given good training and new insights to the students
and their teacher alike.

Other Investigations

Stories comparable to those just presented could be told about other
relatively small sets of investigations. Many such projects were undertaken to explore the factors that contribute to the exemplary efficiency
of enzyme reactions. The endeavors encompassed regiospecific conversions of molecules oriented about metal ions, inducing very efficient
reactions at the *ortho* positions with phenols and anilines (Brackman,[254–259] Engelsma[260,261]).

Investigations on covalent catalysis also come back to mind. Its
reality could be directly proven at an early date in the case of the
imidazole-catalyzed hydrolysis of nitrophenyl acetate and phenyl
oxalate. Brouwer and Van der Vlugt,[262,263] through quantitative spectroscopic investigation, established the occurrence of the unstable acyl-
imidazole as the intermediate in the hydrolysis. Scrutiny of the role of
the imidazole groups of ribonuclease A constituted an indirect follow-
up to these first modest explorations of the mechanistic features of
enzymatic catalysis.

I apologize to my co-workers because, for reasons of space and
time available, many investigations of shorter lines of research could not
be dealt with in this overview. In my memory—and I trust that the
same is true for the students and *doctorandi*—the investigations concerned have been equal in enjoyment and significance to those of the
larger series. Subjects have always been chosen on the basis of the
preference of the researcher and on the basis of what was anticipated, at
that time, to bring about the most valuable advancement of science.

Peptides and Proteins

Just before the war, in 1939, I got the opportunity to start working in
the field of protein chemistry. I was asked by my mentor, F. Kögl, to
join the investigations centered around the hypothesis that one of the
basic aberrations connected to uncontrolled tissue growth and to
development of cancer might be a decrease in the stereospecificities that
are characteristic of normal metabolic reactions. This could manifest

itself by a partial racemization of certain amino acids, constituents of peptides and proteins.[264]

As is well known by now, this hypothesis—stimulating to interesting research as it may have been—was not supported by subsequent experimental tests at various research centers. In all probability, the initial reports about isolation of partially racemic glutamic acid from tumor tissue were incorrect. My own modest effort, consisting of the search in tumor tissue for enzymes that were adjusted to react with peptides containing D-amino acid residues, gave a negative answer. Even so, I had experienced an interesting introduction into the world of protein chemistry. This interest continued with a different orientation during wartime, in the course of medical chemical research at the Veterinary Faculty at Utrecht.

Like many other chemists, I became more and more impressed by the wonderful structures and architecture of protein molecules. This appreciation remained throughout my education as a chemist. It is somewhat strange to me that—whereas the history of protein chemistry contains the names of Emil Fischer, Max Bergman, Leonidas Zervas, and many other great scientists—the younger generation of organic chemists does not focus much effort on the study of nature's most impressive molecules. If this is a symptom of a general tendency among organic chemists to retreat from fields where chemists of other disciplines are working and to narrow the field of organic chemistry, it should become a subject of consideration and possibly of some concern. It seems that the number of organic chemists also tends to be low in other multidisciplinary fields like metal–organic compounds and homogeneous catalysis. This is a disadvantage to the development of both the multidisciplinary subjects and the organic disciplines. The pendulum will probably sway back in due time, as it did in photochemistry, where chemists of various backgrounds and expertise enjoy fruitful cooperation. Certainly this would be to the benefit of organic chemistry, also.

At Leiden the long-term study of peptides and proteins was put on the program after the war. In the course of the following 40 years, no less than 25 students prepared their theses based on experimentally demanding research in this field. The first was E. S. Bruigom,[265] whose thesis appeared in 1950. The synthetic efforts at that time did not extend beyond tripeptides and their tyrosyl residues, which were known to fulfil important functions in the biological reactions of peptides and proteins. In 1985 the thesis of J. Serdijn[266] was finished. Serdijn was the last of a series of six *promovendi* who worked on the detailed structure of the active center and the mode of action of the enzyme ribonuclease A. Significantly, in the latter investigations the syntheses of stereochemically pure polypeptides with some 20 amino acid residues were considered normal, although still not easy, accomplishments. Today's possibilities of synthesizing even considerably larger polypeptides are discussed in *The Concept and Development of Solid-Phase Peptide Synthesis* written by Bruce Merrifield.[288]

I have made a curious observation about the admittedly restricted number of students who in recent years chose to do their advanced course work in the field of peptide synthesis. Students with little experience now see the preparation of a polypeptide of considerable length and complexity as a difficult but not excessive challenge. They feel convinced that they are going to succeed, and in fact they generally prove to be successful. Some 30 years ago many experienced synthetic chemists would have felt burdened by the task of making an optically pure pentapeptide. This phenomenon resembles the amazing confidence of young children (in the Netherlands) who without hesitation start riding a bicycle right away, whereas my friends and I, in our time, received instruction in this activity and nevertheless made a few painful landings on the ground.

To realize the synthesis of a tricky polypeptide, expert guidance remains as decisive today as in former times. The peptide and protein group at Leiden had excellent research leaders, first K. E. T. Kerling for a long period, and later W. Bloemhoff. Moreover, C. Schattenkerk worked there during the whole period. She earned the admiration and

Mrs. C. Schattenkerk.

thanks of many workers in and outside the peptide group for her outstanding capacities as an experimental researcher and as an invaluable advisor to students and staff members alike.

Polypeptide synthesis, however important, was practiced in our program not so much in its own right as to construct the desired variations in molecular constitution and geometry. These served the purpose of investigating the relationship between detailed molecular structure and chemical or biological properties. Our interests were more in *natura artis magistra* (that is, in studying and learning from the amazing chemistry of the living cell) than in studying biology at the molecular level.

In the 1960s a rather elaborate study was made of angiotensin II, an extremely active blood-pressure-increasing octapeptide. It originates by enzymatic fission of the decapeptide angiotensin I, which in its turn is formed by degradation of still larger peptides. Discovered in 1954 by the group of Skeggs,[267] this angiotensin II has the structure: asparagyl–arginyl–valyl–tyrosyl–isoleucyl–histidyl–prolyl–phenylalanine. Important investigations on the structure–activity relationship were performed by the groups of Page and Bumpus[268] and of Schwyzer.[269]

If we take away residue 1 (Asp), the blood-pressure-increasing activity is diminished but remains very large. However, the activity is almost completely lost upon removal of the arginyl residue at position 2. At Leiden a series of heptapeptides were synthesized in which residue 2 was systematically varied.[270] From the study of these analogues, conclusions could be drawn with regard to which properties of the arginyl residue are essential to its function.[271] An interesting effect that may have practical application is seen upon substituting the L-amino acid residue by its D antipode. This often results in a significant increase of pharmacological activity that may result partly from better resistance against stereospecific degrading enzymes. In addition to residue 2 (Arg) of angiotensin II, residues 6 (His) and 8 (Phe) were also varied.[272,273] It seems as though the phenylalanyl residue at position 8 in particular is responsible for the stimulating influence at the receptor site and that the presence of an aromatic ring as well as exact steric fitting are required. Significantly, substitution of L-thienylalanyl at position 8 gives an analogue with considerable (60%) activity, whereas introduction of the D-thienylalanyl residue at this place leads to inactivity.[273]

This type of investigation into the relationship between chemical structure and pharmacological activity of polypeptides understandably proves to be stimulating to the organic chemist and is also interesting with respect to possible medical applications. Nevertheless, the present stage resembles working partly in the dark, because no exact data are known on the reaction partner, the receptor with which the polypeptide interacts. Structural knowledge of the receptor sites will probably

Angiotensin II.

asparagyl[1] — arginyl[2] — valyl[3] — tyrosyl[4] — isoleucyl[5] — histidyl[6] — prolyl[7] — phenylalanine[8]

become available in the near future, but at this moment, the number of unknowns in the chemical mechanism of the action of many hormones makes the studies rather diffuse. The study of the mechanism of enzymatic processes appears more promising. Quantitative measurements of rate constants are possible, and well-defined intermediates can be traced. Thanks to the progress of polypeptide synthesis, both reaction partners—enzyme and substrate—can now be studied effectively.

In the past enzyme reactions suffered from a prohibitive difficulty with regard to mechanistic scrutiny. A reaction mechanism cannot easily be established in a direct way. The ultimate criterion to decide between the various possible mechanisms of a certain reaction often is the result of systematic variation of the molecular structure of the reaction partners. A classical example is the Hammett ρ value for aromatic reactions that indicates the nature of the reaction on the basis of its sensitivity to electron donation or withdrawal at the reaction center of the aromatic partner. More generally, the nucleophilicity (or electrophilicity) of a reaction step can be deduced from the influence of electron-donating and electron-attracting substituents. To establish, for example, the steric features of a reaction, the logical approach is to evaluate the steric requirements of both reaction partners by using the influence of substituents with well-known steric characteristics.

With enzymatic processes, valuable insight could be gained at an early stage by variation of the (mostly low-molecular-weight) substrate molecule. But the complementary variation of the active site of the enzyme remained too difficult from a synthetic point of view for a long time. However, with the growing potential of synthesis by both classical and solid-phase methods, the balance swayed to the positive side.

Enzymatic reactions have several advantages for mechanistic studies. As a rule they follow one well-defined pathway, with a unique mutual orientation of the reaction partners. Moreover, valuable indications of this arrangement can be obtained in many cases by X-ray analysis. This compares favorably with "normal" reactions in solution of small or medium-size molecules.

Investigations were undertaken at Leiden in 1965–1975 to study homogeneously catalyzed peptide bond formation (Provó Kluit,[274] Kolen[275]). The reaction in acetonitrile of benzoyloxycarbonyl-L-alanyl cyanomethyl ester with benzylamine, catalyzed by bases like imidazole and pyrazole (Kerling, Kolen[275]) was chosen as an appropriate model. Its kinetics can be measured conveniently because the reaction effects a considerable change in optical rotation. To our surprise, no less than five mechanistically different pathways appeared to be followed in this "simple" homogeneous reaction. Taking all these aspects together, the mechanistic study of enzymatic processes at this stage has unique and attractive features to offer the classical organic chemist who likes to learn about the detailed mechanism of optimized chemical reactions.

Many aspects of peptide and protein chemistry will be dealt with in the Merrifield volume.[288] Therefore, I will restrict myself to indicating a few singular results obtained at Leiden with a study that used analogues of pancreas ribonuclease.

Why did we choose RNase[276] in the mid-1960s? It is a stable compound that at an early stage had been isolated in a pure form, crystallized, sequenced, and thoroughly studied by X-ray analysis and other methods.[277,278] Moreover, the molecules consist of a single peptide chain of 124 residues (see structure) that can be split smoothly, between residues 20 and 21 in particular, to yield two moieties (Scheme XIV). Both the smaller part (S-peptide) and the larger part (S-protein) are enzymatically inactive. However, when brought together they bind via noncovalent interactions to form a complex (RNase S) that has complete activity. If residues 15–20 are removed from the S-peptide, the remaining 1–14 sequence still is able to form a 100% active complex with the S-protein. Generally, if a change or substitution is made at positions 1–14, the capacity to form an active complex may be diminished but need not be destroyed. As can be established in separate experiments, many of these residues have a function in the binding of the S-protein or of the substrate. However, if the residue at position 12 (His) is varied, activity is lost completely. From these and other data obtained in various research centers, the conclusion has been reached that His-12 is directly engaged in the chemical reaction of the enzyme and the substrate.

```
 1    2    3    4    5    6    7    8    9   10   11   12   13   14   15   16   17   18   19   20
Lys-Glu-Thr-Ala-Ala-Ala-Lys-Phe-Glu-Arg-Glu-His-Met-Asp-Ser-Ser-Thr-Ser-Ala-Ala

21   22   23   24   25   26   27   28   29   30   31   32   33   34   35   36   37   38   39   40
Ser-Ser-Ser-Asn-Tyr-Cys-Asn-Gln-Met-Met-Lys-Ser-Arg-Asn-Leu-Thr-Lys-Asp-Arg-Cys

41   42   43   44   45   46   47   48   49   50   51   52   53   54   55   56   57   58   59   60
Lys-Pro-Val-Asn-Thr-Phe-Val-His-Glu-Ser-Leu-Ala-Asp-Val-Gln-Ala-Val-Cys-Ser-Gln

61   62   63   64   65   66   67   68   69   70   71   72   73   74   75   76   77   78   79   80
Lys-Asn-Val-Ala-Cys-Lys-Asn-Gly-Gln-Thr-Asn-Cys-Tyr-Gln-Ser-Tyr-Ser-Thr-Met-Ser

81   82   83   84   85   86   87   88   89   90   91   92   93   94   95   96   97   98   99  100
Ile-Thr-Asp-Cys-Arg-Glu-Thr-Gly-Ser-Ser-Lys-Tyr-Pro-Asn-Cys-Ala-Tyr-Lys-Thr-Thr

101  102  103  104  105  106  107  108  109  110  111  112  113  114  115  116  117  118  119  120
Gln-Ala-Asn-Lys-His-Ile-Ile-Val-Ala-Cys-Glu-Gly-Asn-Pro-Tyr-Val-Pro-Val-His-Phe

121  122  123  124
Asp-Ala-Ser-Val
```

Amino acid sequence of ribonuclease A.

A similar situation exists at the end of the RNase polypeptide chain. Merrifield and his colleagues demonstrated that the tail (residues 114–124) can be removed and the activity then can be restored by adding an end peptide again.[279] For good (noncovalent) binding, this substituted tail has to have some extra length (residues 108–124). Here an essential function in the reaction was established for His-119.

In the three-dimensional structure, as determined by X-ray analysis, residues 119 and 12 appear to be situated near each other, in

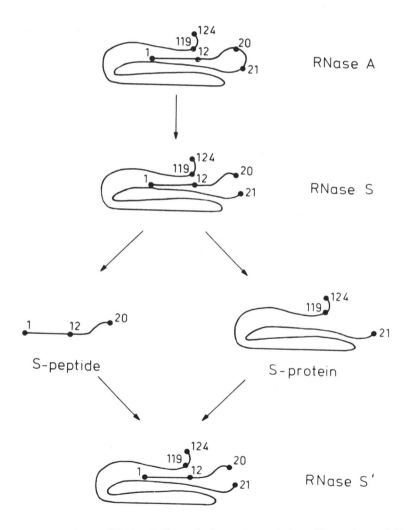

Scheme XIV. "Reversible" splitting of ribonuclease A into S-protein and S-peptide.

the molecular cleft where the substrate is bound and the reaction takes place.[277,278]

The question then arises as to what specific functions these two essential His residues may have. One possibility, of course, is that they may function as proton donors or acceptors in the hydrolysis of the phosphate ester bond (Scheme XV).

Such a description received strong support when it was found that replacing the His residues at position 12 or 119 by a perfectly fitting isosteric residue of different pK, such as L-3-pyrazolylalanyl or L-4-fluorohistidyl residue, results in good substrate binding but practically zero enzymatic activity.[266,280–282,284] At Leiden, as an alternative, the His residues concerned were replaced by residues of similar pK values.[283] For the first time, RNase analogues were obtained that showed good enzymatic activity, even though an essential reacting residue of the active site had been replaced.

For residue 119 the most significant result was obtained by substitution with the N^π-methyl and the N^τ-methyl-L-histidyl residues.[266,284] Significantly, only the N^τ-methyl analogue with an unsubstituted N^π atom shows enzymatic activity. Its isomeric sister with the methylated N^π atom is inactive.

The primary conclusion is that not every alteration of a reacting group in the active center of an enzyme necessarily leads to inactivation. Evidently one of the requirements for activity with residue 119 of RNase is that the group has acid–base properties with a pK of about 6. It appears that of two residues so similar as N^π-methyl and N^τ-

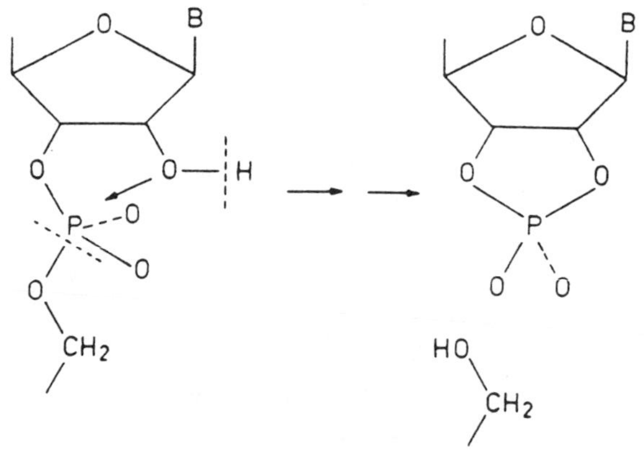

Scheme XV. Hydrolysis of phosphate ester bond in a ribonucleotide.

methylhistidyl, one isomer has activity and the other has not. Apparently, in the genuine enzyme the *pros* nitrogen of His-119 exerts a catalytic function, probably by acting as a proton donor–acceptor in the reaction.

Then what could be the function of the other nitrogen atom of the imidazolyl residue at position 119? The quantitative data on complex formation and reaction rates suggest that it may help somewhat by keeping the imidazole ring in the appropriate orientation. However, the main function should not be overlooked: the one nitrogen atom assures that the other nitrogen of the imidazole ring has a pK value that is appropriate for acting as an efficient acid–base catalyst in the enzymatic process.

With regard to the residue at position 12, one of the approaches made was unorthodox in enzyme chemistry: substitution of the imidazolyl residue by a pyridyl base with the expectedly correct pK but with only one nitrogen in the ring.[285,286] Again the results proved to be thought-provoking. The 2-pyridyl and the 3-pyridyl analogues showed no activity; the 4-pyridyl isomer proved to be highly active. The results show that in the native enzyme the histidine residue at position 12 effects an acid–base catalysis through the *tele* nitrogen. The many details observed and the illuminating NMR results cannot be dealt with here, but it is clear that this time a rather crucial variation of the reacting center itself could result in an enzymatically active complex!

The data can be rationalized on the basis of a hydrolysis mechanism of the phosphate ester bond, as indicated in Scheme XVI. The two

pros(π)- *and* tele(τ)-*methylhistidine residues.*

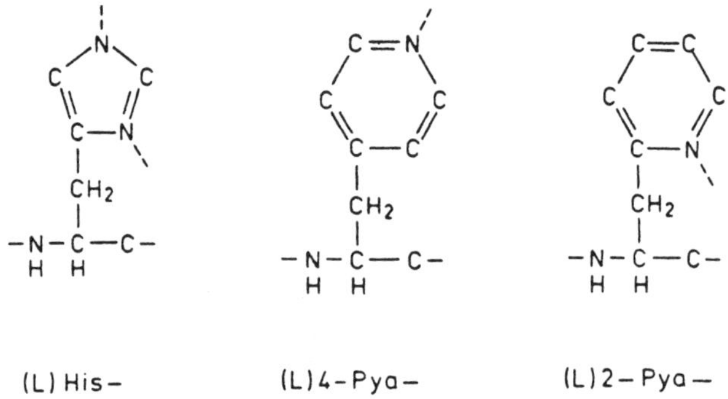

L-Histidine, L-4-pyridylalanine, and L-2-pyridylalanine residues.

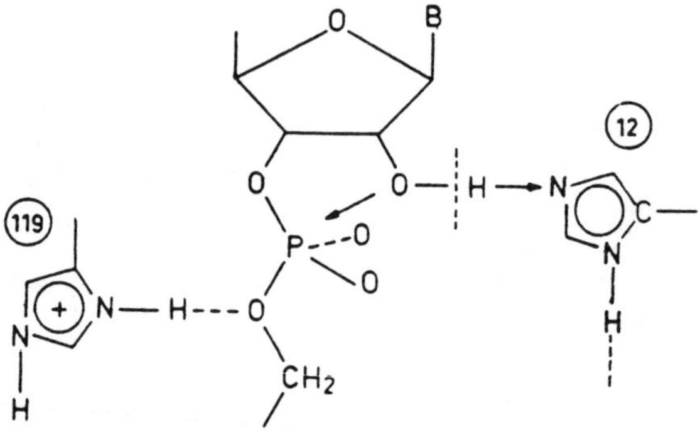

Scheme XVI. Mechanism of the splitting of a ribonucleotide by ribonuclease.

steps of the process are catalyzed by the proton-exchange functions of the two histidine residues at positions 12 and 119. Moreover, the specificities observed with the isomeric analogues indicate that with residue 12 it is the *tele* nitrogen and with residue 119 it is the *pros* nitrogen that directly takes part in the reaction process. The picture has become sufficiently detailed, as an organic chemist likes to have it, for quantitative understanding and further elaboration. On the one hand, there is evidently room for modifications in the construction of the active center. On the other hand, a small difference in the total organization, such as a 3-pyridyl isomer instead of a 4-pyridyl group, results in a dramatic effect on what seems to be an (admittedly sophisticated) acid–base-catalyzed ester hydrolysis.

Of course, all this represents just a beginning. A next step should be an X-ray study of the pyridyl analogues, the inactive 2-pyridyl and 3-pyridyl isomers, and the active 4-pyridyl isomer. Continuation of this line of investigation by using a combination of systematically planned variation of molecular structure—including isotopic substitution—and the diverse physical methods available will lead to detailed understanding of what is essential to optimum catalysis. One of the results will be the synthesis of enzyme analogues or, more generally, of very efficient catalysts with stabilities and specificities different from what is found in nature. From the synthetic point of view, it is encouraging that large fragments of a protein molecule often can be recombined by covalent bonding (protein semisynthesis)[287] and that the de novo synthesis of complete proteins appears feasible, with solid-phase techniques as well as coupling methods in solution.[288,289]

Aims and tasks in this area are certainly demanding. There is every reason for organic chemists to join the research in this and similar fields. The organic chemist is able to design and synthesize the sophisticated molecules that are needed and has a genuine interest in the study of reaction mechanisms. This interest is shared with colleagues from other disciplines.

Organic chemists should not forget the adage, *artis natura magistra*. Learning from the way nature has optimized chemical reactions, the basis of life, and transposing this insight into new processes, novel compounds, and theoretical understanding are among the most important goals of our science.

It seems probable that, provided chemistry is given sufficient means and time, the adept will make alternatives to the systems of nature that are mightier than what has evolved during billions of years. Let us hope that *Homo sapiens* by that time will have adapted enough to use this unique position and potential wisely, and thus continue to enjoy the *Zauberlehrling* of Goethe and Dukas's sublime musical allegory.

Learning and Building

The preceding sections contain sketches of a few main lines of research that I have been directly engaged with during the past 50 years. I thought it appropriate to complement these rather specialized reports by indicating some aspects of the development of organic chemistry at Leiden and at other places, interaction with colleagues and students, and "parachemical" activities like the building of laboratories at Utrecht and at Leiden. I hope that these fragmentary remarks may help to form an impression of the evolution of chemistry and life at a European university during a good part of the 20th century.

Education, Students

The most direct and strongest influence on an organic chemistry professor comes from the students who largely accomplish their teacher's synthesis and continuous adaptation to new developments. I have been privileged to guide the education of a large number of students and to have been educated by the steady stream of these same students.

Organic chemistry, by its cultivation of experimental research, fosters person-to-person teaching, close cooperation in groups, and the daily exchange of experiences and ideas. Often in this field of science one has to digest failures, to overcome barriers, and to make a new start. As a compensation, there is the stimulus of daily encouragement from junior and senior colleagues at the bench.

Theoretical and experimental organic chemistry.

My friend Luut Oosterhoff, himself a theoretician but very appreciative of experimental work, observed that many organic chemistry teachers are benevolent despots. He naturally produced a hypothesis and related this attitude to the demanding mastery this species has to have over the huge collection of facts and phenomena to be handled in organic chemistry. On the other hand, this same organic chemistry by its nature leads to democratic relationships. Whether a first-year student or a venerable professor makes a suggestion or prediction does not count so much; the experimental follow-up is weighed, and the information and answers are accepted by all parties.

At the risk of being considered seriously biased, I would like to report that although only a small number of women choose to study organic chemistry, this in my experience is compensated by the excellent capacities of these few in both experimental research and theoretical exploration.

I served at the university while lectures and oral examinations were still fashionable. These experiences catalyzed personal acquaintance and sometimes mutual appreciation between students and teachers. Such was certainly the case with the senior students. A substantial percentage chose to continue after their final examination (in the Netherlands, *doctoraal examen*).

The final examination is followed by 4 years or even longer of intensive research on a subject that is carefully chosen by student and mentor together. Eventually the student is ready to write a thesis, defend it, and earn the doctorate. The public defence against the opposition of the faculty and before an audience of family, colleagues, and friends has a ceremonial character. The doctoral candidate is assisted by two paranymphs, whose special tasks formerly included practical assistance in case of a physical fight aroused by the passionate dispute. Let me add that contemporary discussions on the contents of the thesis as a rule are friendly but, nevertheless, real and substantial. The day always ends in a glorious traditional fashion with speeches, a reception, and appropriate festivities.

The most complete report on my activities in organic chemistry at Leiden is to be found in the series of the 166 doctoral theses written and defended under my supervision. Moreover, a scientific family relationship develops between the former students who have worked in the same laboratory, often on related subjects, and their teachers and supervisors. This may be happening somewhat less with the younger generation and with the larger number of students. But as far as I can tell, it is still there, and it constitutes one of the most valuable aspects of being and remaining an adept of the university in the broad sense of the word.

Evidently an education in the field of organic chemistry provides a basis for a multitude of amazingly varied careers. Many of our stu-

Senate Room of Leiden University. The Board of Examiners are about to award a doctorate. From left to right, Havinga, C. Altona, A. A. H. Kassenaar (Rector Magnificus), J. H. van Boom, J. Lugtenburg, J. J. C. Mulder (face only), and A. van der Gen.

dents became directors of research at widely different industries. Others proved their capacities in governmental and political top functions, in administration, as business managers, and last but not least, in teaching and research at universities in and outside the Netherlands. I feel particularly grateful that former students and now colleagues cultivate a faculty of chemistry at Leiden characterized by stimulating interaction and cooperation.

Colleagues

Cooperation with colleagues and staff members is almost as important as interaction with students. Until 1945 the Leiden chemistry faculty had a physically closed iron door, and an intellectual iron curtain existed between the inorganic and organic chemistry departments. After the war inorganic and organic chemistry—and later physical chemistry and biochemistry—started to function as parts of one and the same science. Because relationships with other fields of science had been good

North Sea Beach, in the south of the Netherlands, 1980. Enjoying a walk with J. H. Oort, Professor Emeritus of Astronomy at Leiden. Born in 1900, Oort is still active in science.

all the time, scientific and social interactions at Leiden during the years 1946–1970 were fine and stimulating.

In the late 1960s the wave of democratization hit the University of Leiden, like so many other centers of teaching and research all over the world. It did not deeply affect the chemistry department, where attention is strongly focused on experimental research. More insidious has been the gradual infiltration by bureaucracy and overorganization that aims at directing and "improving" scientific activities. This change is good in principle, but effectively causes hindrance and loss of time by overproduction of questionnaires, micropolitical dicussions, continuous reorientations, renovations, etc. I often feel downhearted when I see my younger colleagues, in the potentially most productive period of their life, spending much time and energy on barely effective administrative and organizational business.

Fortunately, organic chemists have been trained to cope with adversities. It now seems as though they have built up resistance against bureaucracy and too-frequent organizational interference. Now (1987) there is an opportunity for organic chemistry and generally for science at the university to adapt to the new boundary conditions and to concentrate on research regarding basic questions at the frontier of science.

Contacts Abroad

For my generation it was an extraordinary experience after the war was over to be able to move freely to other countries, although of course within very narrow budgetary limits. Then I personally met and came to know outstanding scientists like the musically gifted Linderstrøm-Lang and his friend Niels Bohr at Copenhagen, Tiselius and Theorell in Sweden, the renowned antipodes Karrer and Ruzicka at Zürich, Crowfoot-Hodgkin, Robinson, Ingold, and Todd in Great Britain, and Helferich in Germany.

My first visit to the United States was in 1951–1952, when a Rockefeller grant made it possible for me to work for almost a year in California with Pauling and with Calvin. Two events during the start of this period stand out in my memory. Upon arrival at New York, I paid a short visit to Columbia University. The first person I met in America, working at the bench in a long almost-white coat, was Prelog, the most authentic and representative European chemist I could imagine. The

At the Bürgenstock Conference, with V. Prelog.

next station on my route West was the University of Illinois at Urbana, where Mr. and Mrs. Roger Adams gave us an unforgettable cordial welcome.

My wife and I, living on a one-person grant, often had to make the difficult choice between buying gasoline for our faithful $200 Plymouth and having a meal of doughnuts and coffee. However, we managed both to work hard and to travel quite a lot, to see places all over the country, and to meet many colleagues.

The contacts with groups in other countries represented important scientific education and supported research at Leiden. Direct examples are in the field of proteins and peptides, the profit we had from the methods used at the California Institute of Technology (H. Itano and W. A. Schroeder), and the effective isotope techniques (J. Basham and others) learned during my stay as a guest of the photosynthesis group at Berkeley.

Development at Leiden

It does not appear feasible within the scope of this section to enumerate the names and contributions of the very many persons who played a significant role in the history of organic chemistry at Leiden since 1945. A few are selected for obvious reasons.

C. Koningsberger, now professor emeritus at the Technical University at Eindhoven, directly after the war agreed to share with me the load of readjustment of teaching and research. He was nominated as the successor to the well-known lector (reader) of organic chemistry, J. van Alphen. Koningsberger was well known for his outstanding expertise in the organic chemistry of macromolecules. We worked closely together until 1957, when he was nominated as a full professor at Eindhoven. His position at Leiden then was taken by W. Stevens, one of the many excellent pupils of Backer at Groningen. He functioned to the benefit of the organic chemistry department for more than 15 years and accomplished with his group the synthesis of the long-evasive formic acid anhydride as a final contribution.

A characteristic development after the war—correlated with the expansion and diversification of chemistry—consisted of the nomination of professors with new specializations. For more than 70 years there had been just two chairs of chemistry at Leiden, one of inorganic chemistry and one of organic chemistry. The first to be nominated in the latter position, at the same time the first full professor of organic chemistry in the Netherlands (even in Europe), was the influential A. P. N. Franchimont (who served 1874–1914; biography[290]). In 1902 and the following years—while he himself was at Leiden—all other chairs of organic chemistry in the Netherlands were occupied by his pupils: Lobry de Bruyn (rearrangements) at Amsterdam (1896–1904); A. F. Holleman (orientation rules of aromatic substitution) at Groningen (1893–1904) and Amsterdam (1904–1924); and P. van Romburgh (natural products) at Utrecht (1902–1925). Franchimont was succeeded by J. J. Blanksma (1914–1945), who did a lot of physical organic chemistry *avant la lettre* (reaction kinetics).

In the course of my professorship, starting in 1946, this typical European system was gradually expanded. New chairs were created and departments were formed that now count several full professors, senior staff members, and a full-fledged support staff. From the chair of organic chemistry, the chairs of theoretical organic chemistry (L. J. Oosterhoff) and of biochemistry (H. Veldstra) branched off. E. C. Kooyman, a pupil of Wibaut at Amsterdam, was nominated to a new chair of organic chemistry. Before his stay at Leiden, Kooyman had been director of organic and bioorganic research at the central research laboratory of the Shell Corporation at Amsterdam. He did outstanding research in the fields of free radical chemistry, aromatic substitution, and metal–organic compounds. After some 10 years as a professor of organic chemistry, Kooyman accepted a leading position in industry but retained an active connection with Leiden until his death in 1980.

His successor as a full professor was H. Kloosterziel, an adept from the school of Backer at Groningen. He had combined working at the Shell laboratory at Amsterdam with an extraordinary professorship at Eindhoven. Many organic chemists will know his name because of his early investigations on pericyclic reactions, cycloheptatrienes, and conjugated carbanions. Kloosterziel, who as a young student had made exceptionally brave accomplishments in the resistance movement against the Nazis and who was awarded the Medal of Freedom of the United States, died in 1986 after a long illness.

A promising adept from Leiden, who stayed for a longer time at his university, was M. E. Kronenberg. He became well known for his research in the field of photochemistry. After a productive period as a lector of organic chemistry, he chose to follow a managerial career that was cut off by his death in 1977.

Visitors

Very valuable interactions have occurred through the many visits of foreign scientists to Leiden. Among the first was that by Schwyzer from Zürich, who came down the Rhine River with a group of students to see and practice the special techniques of microscale synthetic chemistry, originally developed by my teacher, Kögl, and adopted at Leiden.

The first visitor from the United States to spend a sabbatical (1951–1952) at Leiden was Charles Bradsher (Duke University), who with his wife Dorothy and their children, lived (and managed to survive) in our house while we were in America. They accepted with a good sense of humor the picturesque but comparatively primitive conditions of life in an old Dutch city. The latter difficulty was experienced to a lesser extent by Hans and Elly Wijnberg. It is gratifying that this sabbatical may have contributed to a definite return to our country and to a most successful professorship at the University of Groningen. Another dear visitor of Netherlands origin has been William le Noble (Stony Brook). His and his wife's stay represented one of the highlights of organic chemistry at Leiden. And then we remember the Kwarts, Harold and Helen (University of Delaware), who came to Leiden so often and whose deaths were a great loss also to our department. At Leiden we feel very grateful to all those who came to visit and work at the old laboratory in the Hugo de Grootstraat or later at the Gorlaeus Laboratories. They became part of the laboratory family, and it is always a pleasure to see them visiting our place again and renewing scientific contacts and friendships.

Relation with Industries

In my experience, the relationship between the chemical industry and university has been particularly sound and good in the Netherlands. Quite a few large industrial concerns like Unilever, Shell, Philips, AKZO, DSM, and others have primary research facilities in this country. These centers have given an example of how to cultivate fruitful interaction with university laboratories without interfering with their research program. As a rule, and within sensible limits, there is free exchange of ideas, experiences, and special compounds between university and large industrial laboratories. In my experience, this exchange constitutes a great help and stimulus to university research, certainly in the field of organic chemistry.

The industries, on their side, adhere to the philosophy that regular production of well-trained young chemists is important to them. Moreover, the results obtained and the novel methods developed at the university in this relaxed type of mutual support are accessible to the workers at the industrial laboratories. So-called contract research is advocated as an exception rather than as a normally desirable form of cooperation. The university institutes should primarily concentrate on research in fields selected because they are the most intriguing and promising with regard to deepening insight, developing novel methods of synthesis, and scrutinizing molecular dynamics, the basis of life.

Organic Chemistry and Architecture: Building Laboratories

The longer I have lived with organic chemistry, the more I feel that my drive toward scientific research is related to the appreciation of art and architecture. I started my career in chemistry in a curious old building of the Veterinary Faculty at Utrecht. Later I was asked more than once to combine my interests in chemistry and architecture, more specifically in building laboratories.

Before the war, in the 1930s, a new laboratory had to be constructed at Utrecht. My mentor, Kögl, functioned as a faculty spokesman. As a physically oriented organic assistant, I got the task of giving advice on the required technical facilities, particularly regarding electricity. The building at the Croesestraat, finished in 1939, was of a simple style and construction, different from the imposing facade architecture of older chemical laboratories. But even so, the layout was typical of the prewar European tradition. There was one impressive room for the professor, with a room next to it for his secretary, and a private laboratory where a senior assistant and a few selected co-workers could do their research under the close supervision of the chief. Of course, large rooms were available in which students could work, along with a lecture hall, a library, etc. Characteristically, there was nothing like a faculty room or a cafeteria, and nobody bothered about it.

In these one-professor institutes an experienced co-worker or secretary often acquired a decisive influence. Clearly the system had its risks, but it was free from bureaucracy and very efficient. Because of the restricted size of the building and the small number of inhabitants, everyone knew and daily met everyone else.

When I arrived at Leiden (1945), there still was that same atmosphere and situation. The laboratory of organic chemistry at the Hugo de Grootstraat consisted of a romantic building that, particularly at night, had the appearance of a castle with many towers. Although the basement was completely flooded during the last part of the war, the institute had been kept in shape and going as far as possible with the help of senior students E. M. Petri, P. van der Burg, and E. S. Bruigom.

Tradition at Leiden was mightier still than at Utrecht. All facilities and construction details had respectfully been kept exactly as they had been made under Franchimont's direction in 1900. The hoods were ventilated by the use of gas flames. Upon entering the building, one immediately recognized the smell characteristic of an organic chemistry laboratory, which is euphemistically attributed to a composite mixture with a substantial percentage of "benzaldehyde". Rooms and corridors were painted in a time-honored dark brown. After continued desperate efforts and discussions, the whole facility was repainted in light, predominantly grey, tones. Since that revolutionary change, tradition

Laboratory of Organic Chemistry, University of Leiden, 1901–1968.

has taken its course again. Now grey has been adopted as the preferred color for the institutes of Leiden University.

The number of chemistry students greatly increased. Heavy internal strain resulted from steric hindrance in the building, originally made for a maximum of 12 senior students. Moreover, the construction started to show cracks and instability caused by the incorporation of heavy instruments like mass spectrometers and NMR machines.

The decision to move out of the attractive old city of Leiden was not made easily. No longer would morning lectures be accompanied by the moving music of a street organ, which made it barely possible for the professor to be heard, let alone understood, by his amused audience. And no longer would the impressive power of chemistry be visibly and acoustically demonstrated to the citizens by the dumping of considerable surpluses of sodium in the old canal. However, the chemistry faculty realized at an early stage that the future would lie in new facilities with possibilities for further adaptation and substantial expansion.

The new Gorlaeus Laboratories were built in the years 1961–1968. I was asked by the faculty to incorporate the aspects essential to chemical research and teaching into the complex mixture created by representatives of various ministries, the board of the university, the architects, and the many financial and technical experts. One of the architects, G. Drexhage, appeared to be a real genius at convincing everyone present in a meeting of the merits of a certain plan. Thus he succeeded in getting through the design of the valuable broad corridors on the basis of the curious argument that the width of the corridors hardly influences the cost of building. I succeeded in securing his sympathy by suggesting at the first official meeting of the venerable committee that it would not matter so much whether the buildings became suitable to do chemistry if only they would be of nice appearance and good architectural style.

We had the essential cooperation of members of the faculty, including the physical chemist A. J. Staverman and in particular a very able senior staff member, S. J. Roorda, who was later to become managing director of the complete laboratory complex. Furthermore, an excellent member of the technical staff, Roozendaal, practically always gave the best judgment and advice on the ability of construction and materials to withstand specific aggression by compounds and workers in a chemical laboratory.

This is not the place to go into details of the complex process of planning and building, however remarkable these are to an experimental chemist. It seems worthwhile, though, to mention one strategy that has proven to be extremely useful. At the beginning of the elaborate discussion of the project, we asked to follow a pathway dear to the organic chemist. We were allowed to build an experimental model laboratory in which all facilities to be used in the future buildings were tested, as well as the layout of the laboratories, the type and intensity of the artificial light, the ventilation, etc. We had student classes do their practical courses in the test laboratories made exactly according to the designed plan.

In the beginning the results were terrifying. Serious flooding used to occur because of the undercapacity of the drains, for which the special requirements necessitated by the use of water pumps had been underestimated. The beautifully constructed hoods gave rise to impressive explosions. The ventilation caused strong local drafts that interfered with the use of balances. Very many improvements were made in the layout and construction. Finally, after a year of hard work, trial, and error, the test laboratories proved to be efficient and pleasant workplaces. By following this experimental approach, we saved a lot of money and time in the end. The new buildings appeared to be practically free from the otherwise-common "childhood diseases".

It is difficult not to continue and tell about the comical, curious, and sometimes dramatic phases of the building activities that covered some 8 years. But this volume focuses on organic chemistry, often comparable to architectural and building activities, but on a very different

Gorlaeus Laboratories, University of Leiden.

scale. What has been the impact of these activities on the author's and the staff's production in the chemical field? Of course, it took lots of time and attention, and it presented a handicap in producing publications beyond the regular stream of theses. But I cannot find that this handicap has been dramatic. Stated otherwise, research activities have not been essentially more intense in periods without the extra responsibility for laboratory building than in the years 1960–1970. Of course, students, co-workers, and colleagues may have a different and more valuable opinion. However this may be, the contacts with architects and technical and organizational experts have substantially contributed, possibly not to scientific production, but certainly to my general education.

Epilogue

Looking back on the 60 years during which I enjoyed doing organic chemistry, the overriding impression is that of the amazingly fast evolution, almost revolution, in this field of science. Analysis and synthesis have shown a progress undreamt of before. The use of powerful physical methods, NMR in particular, combined with specific-isotope substitution now enables us to "see" molecules in detail, to directly follow the conformational behavior and the reactions of even complex compounds. The rapidly growing computer facilities and the effective methods of calculating molecular properties bring successful approaches from the theoretical side. Although we are only beginning the quantitative description and prediction of the detailed pathway of the reactions of organic molecules, the consistency of the results obtained through different methods of investigation is already impressive.

One may ask whether in science and in chemistry in particular there is no limit to the growth of knowledge, the capacity for investigation, and the deepening of our insight. A second question is related: Does this evolution make human life more complete and happy? With regard to the latter question, I tend to arrive at an answer by transposing it to whether the development of a child to an adult person makes this person or the community any happier. I am inclined to believe that there is no end indeed to the evolution of science and the study of organic chemistry in particular.

Life is largely based on reactions of organic molecules. By studying in depth the properties and behavior of organic compounds, we are scrutinizing our own nature. This we will continue to do as long as our species exists.

More prosaically, I would deal with another question that is related to the contents of the preceding sections. This question was formulated and answered in a personal way at one of the first photochemistry congresses in the United States (Rochester, 1963). After a few days of stimulating lectures, I attended the banquet in honor of the resigning, highly esteemed Nestor of photochemistry, W. Albert Noyes, Jr. On that occasion he was addressed in a grandiose and moving way by his friend and colleague, Nobel laureate Norrish, who asked the following: Would we—you, Albert, and I—start again this career with all its inherent difficulties, the extremely demanding work, the failures and deceptions, if we now (1963) could make the choice again? Norrish's answer, describing his own feeling, was no, not now; perhaps after another 100 years, but not now.

Without being optimistic or lighthearted at the end of my career as a chemist, if I were asked a similar question, I would give a positive answer. *Tempora mutantur, nos et mutamur in illis*; but the deeply seated curiosity, the drive to investigate and to obtain insight and novel vistas, remains essentially unchanged. Remembering all the troubles, difficulties, and doubts in the course of my work, I still would know of no greater perspective than that of making a new start, entering research at the frontier of knowledge and understanding. This presupposes one primary condition, that my partner for life, Louise Diederika Oversluys, would join me again in such an adventurous future.

Acknowledgments

I wish to express my deep gratitude to H. J. C. Jacobs, whose encouragement and continuous help have been decisive in the realization of the whole of the manuscript. The expert comments of C. Altona, W. Bloemhoff, and J. Cornelisse, who read sections of the text, are thankfully acknowledged. Valuable constructive advice has been given by J. I. Seeman, Series Editor.   Thanks are due to J. J. Pot and A. G. Huigen for the helpful production of an important part of the illustrations and figures. The invaluable help and cooperation that I enjoyed from my wife, Louise D. Oversluys, is indicated in the last sentence of this chapter and reflected in the accompanying picture.

References

1. Havinga, E. Thesis, University of Utrecht, Jan. 1939.

2. De Wael, J. Thesis, University of Utrecht, July 1938.

3. Havinga, E.; De Wael, J. *Recl. Trav. Chim. Pays-Bas* **1937**, *56*, 375.

4. De Wael, J.; Havinga, E. *Recl. Trav. Chim. Pays-Bas* **1940**, *59*, 770.

5. Kögl, F.; Havinga, E. *Recl. Trav. Chim. Pays-Bas* **1940**, *59*, 601.

6. Kögl, F.; Havinga, E. *Recl. Trav. Chim. Pays-Bas* **1940**, *59*, 249, 323.

7. Zeelen, F. J. Thesis, University of Leiden, Dec. 1956.

8. Den Hertog-Polak, M. Thesis, University of Leiden, July 1951.

9. Den Hertog-Polak, M.; Havinga, E. *Recl. Trav. Chim. Pays-Bas* **1952**, *71*, 64.

10. Havinga, E. In *Monomolecular Layers*; Sobotka, H., Ed.; American Association for the Advancement of Science: Washington, DC, 1954; p 192.

11. Van Deenen, L. L. M.; Houtsmuller, U. M. T.; De Haas, G. H.; Mulder, E. *J. Pharm. Pharmacol.* **1962**, *14*, 429, and subsequent papers.

12. Havinga, E. *Chem. Weekbl.* **1941**, *38*, 642.

13. Havinga, E. *Biochim. Biophys. Acta* **1954**, *13*, 171.

14. Böeseken, J. *Recl. Trav. Chim. Pays-Bas* **1921**, *40*, 552, 578.

15. Derx, H. G. *Recl. Trav. Chim. Pays-Bas* **1922**, *41*, 312.

16. Hermans, P. H. *Z. Phys. Chem.* **1924**, *113*, 337.

17. Oosterhoff, L. J. Thesis, University of Leiden, Nov. 1949.

18. Hazebroek, P.; Oosterhoff, L. J. *Discuss. Faraday Soc.* **1951**, *10*, 87.

19. Hassel, O.; Viervoll, H. *Acta Chem. Scand.* **1947**, *1*, 149.

20. Dallinga, G. Thesis, University of Leiden, March 1951.

21. Kwestroo, W. Thesis, University of Leiden, Nov. 1954.

22. Kwestroo, W.; Meijer, F. A.; Havinga, E. *Recl. Trav. Chim. Pays-Bas* **1954**, *73*, 717.

23. Tulinskie, A.; DiGiacomo, A.; Smyth, C. P. *J. Am. Chem. Soc.* **1953**, *75*, 3552.

24. Kozima, K.; Yoshino, T. *J. Am. Chem. Soc.* **1953,** *75,* 166.

25. Kozima, K.; Sakashita, K.; Maeda, S. *J. Am. Chem. Soc.* **1954,** *76,* 1965.

26. Bender, P.; Flowers, D. L.; Goering, H. L. *J. Am. Chem. Soc.* **1955,** 77, 3462.

27. Van der Linden, J. A. Thesis, University of Leiden, Sept. 1958.

28. Wessels, E. C. Thesis, University of Leiden, Jan. 1960.

29. Van Dort, H. M. Thesis, University of Leiden, July 1963.

30. Van Dort, H. M.; Sekuur, T. J. *Tetrahedron Lett.* **1963,** 1301.

31. Van Dort, H. M.; Havinga, E. *Proc. K. Ned. Akad. Wet., Ser. B* **1963,** *66,* 45.

32. Geise, H. J. V. H. Thesis, University of Leiden, March 1964.

33. Geise, H. J.; Altona, C.; Romers, C. *Tetrahedron* **1967,** *23,* 439.

34. Geise, H. J.; Tieleman, A.; Havinga, E. *Tetrahedron* **1966,** 22, 183.

35. Barton, D. H. R. *Experientia* **1950,** *6,* 316.

36. Altona, C.; Romers, C.; Havinga, E. *Tetrahedron Lett.* **1959,** (10), 16.

37. Altona, C. Thesis, University of Leiden, March 1964.

38. Klapwijk, T. Thesis, University of Leiden, Nov. 1958.

39. Van der Linde, J. Thesis, University of Leiden, June 1963.

40. Van der Linde, J.; Havinga, E. *Recl. Trav. Chim. Pays-Bas* **1965,** *84,* 1047.

41. Hageman, H. J. Thesis, University of Leiden, June 1965.

42. Hageman, H. J.; Havinga, E. *Tetrahedron* **1966,** 22, 2271.

43. Altona, C.; Hageman, H. J.; Havinga, E. *Spectrochim. Acta, Part A* **1968,** *24,* 633.

44. Hageman, H. J.; Havinga, E. *Recl. Trav. Chim. Pays-Bas* **1966,** *85,* 1141.

45. Adriaanse, A. Thesis, University of Leiden, Dec. 1967.

46. Sikkema, D. J. Thesis, University of Leiden, Sept. 1969.

47. Edward, J. T. *Chem. Ind.* **1955,** 1102.

48. Lemieux, R. U. In *Molecular Rearrangements*; De Mayo, P., Ed.; Wiley: New York, 1964; p 709.

49. Van Woerden, H. F. Thesis, University of Leiden, March 1964.

50. Van Woerden, H. F.; Havinga, E. *Recl. Trav. Chim. Pays-Bas* **1967**, *86*, 341.

51. Van Woerden, H. F.; Havinga, E. *Recl. Trav. Chim. Pays-Bas* **1967**, *86*, 353.

52. Kalff, H. T. Thesis, University of Leiden, July 1964.

53. Kalff, H. T.; Havinga, E. *Recl. Trav. Chim. Pays-Bas* **1962**, *81*, 282.

54. De Wolf, N.; Romers, C.; Altona, C. *Acta Crystallogr.* **1967**, *22*, 715.

55. De Wolf, N.; Henniger, P. W.; Havinga, E. *Recl. Trav. Chim. Pays-Bas* **1967**, *86*, 1227.

56. Planje, M. C. Thesis, University of Leiden, Jan. 1966.

57. De Hoog, A. J. Thesis, University of Leiden, March 1971.

58. De Hoog, A. J.; Buys, H. R.; Altona, C.; Havinga, E. *Tetrahedron* **1969**, *25*, 3365.

59. Romers, C.; Altona, C.; Buys, H. R.; Havinga, E. In *Topics in Stereochemistry*; Eliel, E. L.; Allinger, N. L., Eds.; Wiley: New York, 1969; Vol. 4, p 39.

60. Mossel, A. Thesis, University of Leiden, Nov. 1963.

61. Mossel, A.; Romers, C.; Havinga, E. *Tetrahedron Lett.* **1963**, 1247.

62. Mossel, A.; Romers, C. *Acta Crystallogr.* **1964**, *17*, 1217.

63. Henniger, P. W.; Wapenaar, E.; Havinga, E. *Recl. Trav. Chim. Pays-Bas* **1962**, *81*, 1053.

64. Henniger, P. W. Thesis, University of Leiden, Oct. 1967.

65. Henniger, P. W.; Dukker, L. J.; Havinga, E. *Recl. Trav. Chim. Pays-Bas* **1966**, *85*, 1177.

66. Flapper, W. M. J. Thesis, University of Leiden, April 1976.

67. Romers, C.; Altona, C.; Jacobs, H. J. C.; De Graaff, R. A. G. In *Terpenoids and Steroids*; Overton, K. H., Senior Reporter; Specialist Periodical Report; The Chemical Society: London, 1974; Vol. 4, p 531.

68. Pitzer, K. S. *Science (Washington, D.C.)* **1945**, *101*, 672.

69. Pitzer, K. S.; Donath, W. E. *J. Am. Chem. Soc.* **1959**, *81*, 3213.

70. Buys, H. R. Thesis, University of Leiden, Feb. 1968.

71. Altona, C.; Buys, H. R.; Hageman, H. J.; Havinga, E. *Tetrahedron* **1967**, *23*, 2265.

72. Altona, C.; Hirschmann, H. *Tetrahedron* **1970**, *26*, 2173.

73. Altona, C.; Sundaralingam, M. *J. Am. Chem. Soc.* **1974**, *94*, 8205.

74. Altona, C.; Hartel, A. J.; Olsthoorn, C. S. M.; De Leeuw, H. P. M.; Haasnoot, C. A. G. In *The Jerusalem Symposia on Quantum Chemistry and Biochemistry*; Pullman, B., Ed.; Reidel: Dordrecht, 1978; Vol. 11, p 87.

75. Lugtenburg, J.; Havinga, E. *Tetrahedron Lett.* **1969**, 239.

76. Lugtenburg, J. Thesis, University of Leiden, April 1970.

77. Orbons, L. P. M.; Van Beuzekom, A. A.; Altona, C. *J. Biomol. Struct. Dyn.* **1987**, *4*, 965; Orbons, L. P. M. Thesis, University of Leiden, May 1987.

78. Karplus, M. *J. Chem. Phys.* **1959**, *30*, 11; *J. Am. Chem. Soc.* **1963**, *85*, 2870.

79. Haasnoot, C. A. G.; De Leeuw, F. A. A. M.; Altona, C. *Tetrahedron* **1980**, *36*, 2783.

80. Karplus, M. *Ber. Bunsenges. Phys. Chem.* **1982**, *86*, 386.

81. Eliel, E. L. *Stereochemistry of Carbon Compounds*; McGraw-Hill: New York, 1962.

82. Eliel, E. L.; Allinger, N. L.; Angyal, N. J.; Morrison, G. A. *Conformational Analysis*; Wiley: New York, 1965.

83. Hanack, M. *Conformation Theory*; Academic: New York, 1965.

84. Mulder, J. J. C. Thesis, University of Leiden, June 1970.

85. Reerink, E. H.; Van Wijk, A. *Biochem. J.* **1929**, *23*, 1294.

86. Reerink, E. H.; Van Wijk, A.; Van Niekerk, J. *Chem. Weekbl.* **1932**, *29*, 645.

87. Velluz, L.; Amiard, G.; Petit, A. *Bull. Soc. Chim. Fr.* **1949**, 501.

88. Koevoet, A. L.; Verloop, A.; Havinga, E. *Recl. Trav. Chim. Pays-Bas* **1955**, *74*, 788.

89. Velluz, L.; Amiard, G.; Goffinet, B. *Bull. Soc. Chim. Fr.* **1955**, 1341.

90. Setz, P. *Z. Physiol. Chem.* **1933**, *215*, 785.

91. Van der Vliet, J. Thesis, University of Groningen, June 1946, (first proposition).

92. Havinga, E.; Bots, J. P. L. *Recl. Trav. Chim. Pays-Bas* **1954**, *73*, 393.

93. Bots, J. P. L. Thesis, University of Leiden, Dec. 1952.

94. Van den Bos, B. G. Thesis, University of Leiden, July 1956.

95. Lawson, D. E. M.; Fraser, D.; Kodicek, E.; Morris, H. R.; Williams, D. H. *Nature (London)* **1971**, *230*, 228.

96. Holick, M. F.; Semmler, E. J.; Schnoess, H. K.; DeLuca, H. F. *Science (Washington, D.C.)* **1973**, *180*, 190.

97. Norman, A. W.; Myrtle, J. F.; Midgett, R. J.; Nowicki, H. G. *Science (Washington, D.C.)* **1971**, *173*, 51.

98. Inhoffen, H. H.; Bruckner, K.; Grundel, R.; Quinkert, G. *Chem. Ber.* **1954**, *87*, 1407.

99. Castells, J.; Jones, E. R. H.; Williams, R. W. J.; Meakins, G. D. *J. Chem. Soc.* **1959**, 1159.

100. Verloop, A.; Koevoet, A. L. *Chem. Weekbl.* **1954**, *50*, 803.

101. Havinga, E.; Koevoet, A. L.; Verloop, A. *Recl. Trav. Chim. Pays-Bas* **1955**, *74*, 1230.

102. Rappoldt, M. P.; Keverling Buisman, J. A.; Havinga, E. *Recl. Trav. Chim. Pays-Bas* **1958**, *77*, 227.

103. Koevoet, A. L. Thesis, University of Leiden, Feb. 1956.

104. Verloop, A. Thesis, University of Leiden, Feb. 1956.

105. Rappoldt, M. P. Thesis, University of Leiden, Nov. 1958.

106. Sanders, G. M.; Havinga, E. *Recl. Trav. Chim. Pays-Bas* **1964**, *83*, 665.

107. Havinga, E.; De Kock, R. J.; Rappoldt, M. P. *Tetrahedron* **1960**, *11*, 276, and references quoted.

108. Sanders, G. M. Thesis, University of Leiden, April 1967.

109. De Kock, R. J. Thesis, University of Leiden, April 1959.

110. Havinga, E.; Schlatmann, J. L. M. A. *Tetrahedron* **1961**, *16*, 146.

111. Havinga, E. *Chimia* **1962**, *16*, 145.

112. Schlatmann, J. L. M. A. Thesis, University of Leiden, March 1961.

113. Schlatmann, J. L. M. A.; Pot, J.; Havinga, E. *Recl. Trav. Chim. Pays-Bas* **1964**, *83*, 1173.

114. Pot, J. Thesis, University of Leiden, Nov. 1964.

115. Takken, H. J. Thesis, University of Leiden, Feb. 1971.

116. Havinga, E. *Experientia* **1973**, *29*, 1181.

117. Woodward, R. B.; Hoffmann, R. *J. Am. Chem. Soc.* **1965**, *87*, 395.

118. Longuett-Higgins, H. C.; Abrahamson, E. W. *J. Am. Chem. Soc.* **1965**, *87*, 2045.

119. Fukui, K. *Tetrahedron Lett.* **1965**, 2009.

120. Dewar, M. J. S. *Tetrahedron Suppl.* **1966**, *8*, 75.

121. Zimmerman, H. E. *J. Am. Chem. Soc.* **1966**, *88*, 1564, 1566.

122. Van der Lugt, W. T. A. M.; Oosterhoff, L. J. *Chem. Commun.* **1968**, 1235.

123. Mulder, J. J. C.; Oosterhoff, L. J. *Chem. Commun.* **1970**, 305, 308.

124. Van der Hart, W. J.; Mulder, J. J. C.; Oosterhoff, L. J. *J. Am. Chem. Soc.* **1972**, *94*, 5724.

125. Cf. Seeman, J. I. *Chem. Rev.* **1983**, *83*, 83.

126. Hammond, G. S.; Liu, R. S. *J. Am. Chem. Soc.* **1963**, *85*, 477.

127. Dauben, W. G. *Chem. Weekbl.* **1964**, *60*, 381.

128. Dauben, W. G.; Kellogg, M. S.; Seeman, J. I.; Vietmeyer, N. D.; Wendschuh, P. H. *Pure Appl. Chem.* **1973**, *33*, 197.

129. Baldwin, J. E.; Krueger, S. M. *J. Am. Chem. Soc.* **1969**, *91*, 6444.

130. Courtot, P.; Rumin, R. *Tetrahedron* **1976**, *32*, 441.

131. Vroegop, P. J. Thesis, University of Leiden, Jan. 1972.

132. Vroegop, P. J.; Lugtenburg, J.; Havinga, E. *Tetrahedron* **1973**, *29*, 1393.

133. Gielen, J. W. J.; Jacobs, H. J. C.; Havinga, E. *Tetrahedron Lett.* **1976**, 3751.

134. Jacobs, H. J. C.; Havinga, E. *Adv. Photochem.* **1979**, *11*, 305.

135. Fischer, E. *J. Mol. Struct.* **1982**, *84*, 219.

136. Langkilde, F. W.; Jensen, N.-H.; Wilbrandt, R.; Brouwer, A. M.; Jacobs, H. J. C. *J. Phys. Chem.* **1987**, *91*, 1029.

137. Sanders, G. M.; Pot, J.; Havinga, E. *Fortschr. Chem. Org. Naturst.* **1969**, *27*, 131.

138. Dauben, W. G.; Baumann, P. *Tetrahedron Lett.* **1961**, 565.

139. Dauben, W. G., quoted In Woodward, R. B.; Hoffmann, R. *The Conservation of Orbital Symmetry*; Verlag Chemie: Weinheim, 1970; p 80.

140. Bakker, S. A.; Lugtenburg, J.; Havinga, E. *Recl. Trav. Chim. Pays-Bas* **1972,** *91,* 1459.

141. Okamura, W. H. *Acc. Chem. Res.* **1983,** *16,* 81.

142. Westerhof, P.; Keverling Buisman, J. A. *Recl. Trav. Chim. Pays-Bas* **1956,** *75,* 1243.

143. Sanders, G. M.; Stiefelhagen, J. F. C.; Riemersma, R. A.; Havinga, E., unpublished observations.

144. Boomsma, F. Thesis, University of Leiden, Dec. 1975.

145. Boomsma, F.; Jacobs, H. J. C.; Havinga, E.; Van der Gen, A. *Recl. Trav. Chim. Pays-Bas* **1977,** *96,* 104.

146. Jacobs, H. J. C.; Boomsma, F.; Havinga, E.; Van der Gen, A. *Recl. Trav. Chim. Pays-Bas* **1977,** *96,* 113.

147. Barrett, A. G. M.; Barton, D. H. R.; Russell, R. A.; Widdowson, D. A. *J. Chem. Soc., Perkin Trans. 1,* **1977,** 631.

148. Salem, L. *Science (Washington, D.C.)* **1976,** *191,* 822, and references cited.

149. Maessen, P. A. Thesis, University of Leiden, March 1983.

150. Koolstra, R. B. Thesis, University of Leiden, December 1988.

151. Gielen, J. W. J. Thesis, University of Leiden, Feb. 1981.

152. Jacobs, H. J. C.; Gielen, J. W. J.; Havinga, E. *Tetrahedron Lett.* **1981,** *22,* 4013.

153. Maessen, P. A.; Jacobs, H. J. C.; Cornelisse, J.; Havinga, E. *Angew. Chem.* **1983,** *95,* 752; *Angew. Chem. Int. Ed. Engl.* **1983,** *22,* 718; *Angew. Chem. Suppl.* **1983,** 994.

154. Brouwer, A. M. Thesis, University of Leiden, April 1987.

155. Havinga, E.; De Jongh, R. O.; Dorst, W. *Recl. Trav. Chim. Pays-Bas* **1956,** *75,* 378.

156. Havinga, E. *Versl. Gewone Vergad. Afd. Natuurkunde K. Ned Akad. Wet.* **1961,** *70,* 52.

157. Havinga, E.; De Jongh, R. O. *Bull. Soc. Chim. Belg.* **1962,** *71,* 803.

158. Havinga, E. Proc. 13th Solvay Conf. Chem., Brussels 1965; Wiley: New York, 1967; p 201.

159. Havinga, E.; Kronenberg, M. E. *Pure Appl. Chem.* **1968,** *16,* 137.

160. Cornelisse, J.; Havinga, E. *Chem. Rev.* **1975,** *75,* 353.

161. De Jongh, R. O. Thesis, University of Leiden, Feb. 1965.

162. Stratenus, J. L. Thesis, University of Leiden, Oct. 1966.

163. Kronenberg, M. E.; Van der Heyden, A.; Havinga, E. *Recl. Trav. Chim. Pays-Bas* **1967**, *86*, 254.

164. Lammers, J. G. Thesis, University of Leiden, June 1974.

165. Lammers, J. G.; Tamminga, J. J.; Cornelisse, J.; Havinga, E. *Isr. J. Chem.* **1977**, *16*, 304.

166. Beijersbergen van Henegouwen, G. M. J.; Havinga, E. *Recl. Trav. Chim. Pays-Bas* **1970**, *89*, 907.

167. Van Vliet, A. Thesis, University of Leiden, May 1969.

168. Van Vliet, A.; Cornelisse, J.; Havinga, E. *Recl. Trav. Chim. Pays-Bas* **1969**, *88*, 1339.

169. Van Vliet, A.; Kronenberg, M. E.; Cornelisse, J.; Havinga, E. *Tetrahedron* **1970**, *26*, 1061.

170. Havinga, E.; Cornelisse, J. *Pure Appl. Chem.* **1976**, *47*, 1.

171. Letsinger, R. L.; Ramsay, O. B.; McCain, J. H. *J. Am. Chem. Soc.* **1965**, *87*, 2945.

172. De Vries, S.; Havinga, E. *Recl. Trav. Chim. Pays-Bas* **1965**, *84*, 601.

173. Kronenberg, M. E.; Van der Heyden, A.; Havinga, E. *Recl. Trav. Chim. Pays-Bas* **1966**, *85*, 56.

174. Nijhoff, D. F.; Havinga, E. *Tetrahedron Lett.* **1965**, 4199.

175. Brasem, P.; Lammers, J. G.; Cornelisse, J.; Lugtenburg, J.; Havinga, E. *Tetrahedron Lett.* **1972**, 685.

176. Lok, C. M.; Havinga, E. *Proc. K. Ned. Akad. Wet., Ser. B* **1974**, *77*, 15.

177. El'tsov, A. V.; Kul'bitskaya, O. V.; Frolov, A. N. *J. Org. Chem. USSR Engl. Transl.* **1972**, *8*, 78.

178. Nilsson, S. *Acta Chem. Scand.* **1973**, *27*, 329.

179. Eberson, L. *J. Am. Chem. Soc.* **1967**, *89*, 4669.

180. Cornelisse, J.; De Gunst, G. P.; Havinga, E. *Adv. Phys. Org. Chem.* **1975**, *11*, 225.

181. Döpp, D. O. *Top. Curr. Chem.* **1975**, *55*, 49.

182. Groen, M. B.; Havinga, E. *Mol. Photochem.* **1974**, *6*, 9.

183. Shadid, O. B. Thesis, University of Leiden, Oct. 1979.

184. Den Heijer, J.; Shadid, O. B.; Cornelisse, J.; Havinga, E. *Tetrahedron* **1977**, *33*, 779.

185. Lemmetyinen, H.; Konijnenberg, J.; Cornelisse, J.; Varma, C. A. G. O. *J. Photochem.* **1985**, *30*, 315.

186. Den Heijer, J. Thesis, University of Leiden, May 1977.

187. Kornblum, N.; Michel, R. E.; Kerber, R. C. *J. Am. Chem. Soc.* **1966**, *88*, 5662.

188. Russell, G. A.; Danen, W. C. *J. Am. Chem. Soc.* **1966**, *88*, 5663.

189. Kim, J. K.; Bunnett, J. F. *J. Am. Chem. Soc.* **1970**, *92*, 7463, 7464.

190. Bunnett, J. F. *Acc. Chem. Res.* **1978**, *11*, 413.

191. Miller, G. L.; Miller, M. J.; Crosby, D. G.; Sonturn, G.; Zepp, R. G. *Tetrahedron* **1979**, *35*, 1797.

192. Siegman, J. R.; Houser, J. J. *J. Org. Chem.* **1982**, *47*, 2773.

193. Cornelisse, J.; Havinga, E. *Tetrahedron Lett.* **1966**, 1609.

194. De Gunst, G. P.; Havinga, E. *Tetrahedron* **1973**, *29*, 2167.

195. De Gunst, G. P. Thesis, University of Leiden, Sept. 1971.

196. Tamminga, J. J. Thesis, University of Leiden, April 1979.

197. Varma, C. A. G. O.; Tamminga, J. J.; Cornelisse, J. *J. Chem. Soc., Faraday Trans. 2,* **1982**, *78*, 265.

198. Van Riel, H. C. H. A. Thesis, University of Leiden, Feb. 1977; Havinga, E. *Van't Hoff Herdenkingsrede, KNAW,* 25 November 1978.

199. Van Riel, H. C. H. A.; Lodder, G.; Havinga, E. *J. Am. Chem. Soc.* **1981**, *103*, 7257.

200. Cornelisse, J.; Lodder, G.; Havinga, E. *Rev. Chem. Intermed.* **1979**, *2*, 231.

201. Ferrier, S. J. H. Thesis, University of Leiden, June 1982.

202. Wagner, P. J. *Pure Appl. Chem.* **1977**, *49*, 259.

203. Müller, K. *Angew. Chem.* **1980**, *92*, 1.

204. Veldstra, H. *Enzymologia* **1944**, *11*, 97.

205. Veldstra, H.; Havinga, E. *Recl. Trav. Chim. Pays-Bas* **1943**, *62*, 841.

206. Nivard, R. J. F. Thesis, University of Leiden, April 1951.

207. Havinga, E.; Nivard, R. J. F. *Recl. Trav. Chim. Pays-Bas* **1948**, *67*, 846.

208. Havinga, E.; Nivard, R. J. F. *Recl. Trav. Chim. Pays-Bas* **1949**, *68*, 356.

209. Mattray, O. J. Thesis, University of Leiden, July 1956.

210. Riezebos, G. Thesis, University of Leiden, March 1959.

211. Riezebos, G.; Havinga, E. *Recl. Trav. Chim. Pays-Bas* **1961,** *80,* 446.

212. Laarhoven, W. H. Thesis, University of Leiden, June 1959.

213. Laarhoven, W. H.; Nivard, R. J. F.; Havinga, E. *Recl. Trav. Chim. Pays-Bas* **1960,** *79,* 1153.

214. Laarhoven, W. H.; Nivard, R. J. F.; Havinga, E. *Experientia* **1961,** *17,* 214.

215. Laarhoven, W. H.; Nivard, R. J. F.; Havinga, E. *Recl. Trav. Chim. Pays-Bas* **1961,** *80,* 775.

216. Kronenberg, M. E. Thesis, University of Leiden, July 1962.

217. Kronenberg, M. E.; Havinga, E. *Recl. Trav. Chim. Pays-Bas* **1965,** *84,* 17.

218. Kronenberg, M. E.; Havinga, E. *Recl. Trav. Chim. Pays-Bas* **1965,** *84,* 979.

219. De Vries, M. Thesis, University of Leiden, July 1966.

220. Smith, S. O.; Pardoen, J. A.; Lugtenburg, J.; Mathies, R. A. *J. Phys. Chem.* **1987,** *91,* 804.

221. Smith, S. O.; Palings, I.; Copié, V.; Raleigh, D. P.; Courtin, J.; Pardoen, J. A.; Lugtenburg, J.; Mathies, R. A.; Griffin, R. G. *Biochemistry* **1987,** *26,* 1606.

222. Brouwer, A. M.; Cornelisse, J.; Jacobs, H. J. C. *J. Photochem. Photobiol. Part A* **1988,** *42,* 313.

223. Hammett, L. P. *Physical Organic Chemistry*; McGraw-Hill: New York, 1940.

224. Schors, A. Thesis, University of Leiden, Nov. 1950.

225. Havinga, E.; Schors, A. *Recl. Trav. Chim. Pays-Bas* **1950,** *69,* 457.

226. Havinga, E.; Schors, A. *Recl. Trav. Chim. Pays-Bas* **1951,** *70,* 59.

227. Kraaijeveld, A. Thesis, University of Leiden, Feb. 1953.

228. Kraaijeveld, A.; Havinga, E. *Recl. Trav. Chim. Pays-Bas* **1954,** *73,* 537.

229. Kraaijeveld, A.; Havinga, E. *Recl. Trav. Chim. Pays-Bas* **1954,** *73,* 549.

230. Hodgson, H. H. *J. Chem. Soc.* **1931,** 1494.

231. Anderson, L. C.; Yanke, R. L. *J. Am. Chem. Soc.* **1934,** *56,* 732.

232. Philbrook, G. E.; Getten, T. C. *J. Org. Chem.* **1959,** *24,* 568.

233. Süs, O. *Liebigs Ann. Chem.* **1944,** *556,* 65.

234. De Jonge, J.; Dijkstra, R. *Recl. Trav. Chim. Pays-Bas* 1948, *67*, 328.

235. De Jonge, J.; Dijkstra, R.; Braun, P. B. *Recl. Trav. Chim. Pays-Bas* 1949, *68*, 430.

236. De Jonge, J.; Alink, R. J. H.; Dijkstra, R. *Recl. Trav. Chim. Pays-Bas* 1950, *69*, 1448.

237. Darwin, C.; Crowfoot-Hodgkin, D. *Nature (London)* 1950, *166*, 827.

238. Fenimore, C. P. J. *Am. Chem. Soc.* 1950, *72*, 3226.

239. Hammick, D. L.; New, R. G. A.; Sutton, L. E. *J. Chem. Soc.* 1932, 742.

240. Lüttke, W. *Z. Elektrochem.* 1957, *61*, 302.

241. Gowenlock, B. G.; Trotman, J. *J. Chem. Soc.* 1956, 1670.

242. Gowenlock, G. B.; Lüttke, W. *Q. Rev.* 1958, *12*, 321.

243. Mijs, W. J. Thesis, University of Leiden, March 1959.

244. Schors, A.; Kraaijeveld, A.; Havinga, E. *Recl. Trav. Chim. Pays-Bas* 1955, *74*, 1243.

245. Hoekstra, S. E. Thesis, University of Leiden, Nov. 1960.

246. Mijs, W. J.; Hoekstra, S. E.; Ulmann, R. M.; Havinga, E. *Recl. Trav. Chim. Pays-Bas* 1958, *77*, 746.

247. Havinga, E. *Chem. Weekbl.* 1955, *51*, 125.

248. Umans, A. J. H. Thesis, University of Leiden, April 1959.

249. Fischmann, E. Thesis, University of Leiden, Nov. 1959.

250. Kehrmann, F. *Liebigs Ann. Chem.* 1898, *303*, 1.

251. Romers, C.; Brink Schoemaker, C.; Fischmann, E. *Recl. Trav. Chim. Pays-Bas* 1957, *76*, 490.

252. Romers, C.; Fischmann, E. *Acta Crystallogr.* 1960, *13*, 809.

253. Raban, M.; Greenblatt, J. In *The Chemistry of Amino, Nitroso and Nitro Compounds and Their Derivatives*; Patai, S., Ed.; Wiley: New York, 1982; p 53.

254. Brackman, W. Thesis, University of Leiden, Sept. 1953.

255. Brackman, W.; Havinga, E. *Recl. Trav. Chim. Pays-Bas* 1955, *74*, 937.

256. Brackman, W.; Havinga, E. *Recl. Trav. Chim. Pays-Bas* 1955, *74*, 1021.

257. Brackman, W.; Havinga, E. *Recl. Trav. Chim. Pays-Bas* 1955, *74*, 1070.

258. Brackman, W.; Havinga, E. *Recl. Trav. Chim. Pays-Bas* **1955**, *74*, 1100.

259. Brackman, W.; Havinga, E. *Recl. Trav. Chim. Pays-Bas* **1955**, *74*, 1107.

260. Engelsma, G. Thesis, University of Leiden, June 1959.

261. Engelsma, G.; Havinga, E. *Tetrahedron* **1958**, *2*, 289.

262. Brouwer, D. M. Thesis, University of Leiden, Sept. 1957.

263. Brouwer, D. M.; Van der Vlugt, M. J.; Havinga, E. *Proc. K. Ned. Akad. Wet., Ser. B* **1957**, *60*, 275.

264. Kögl, F.; Erxleben, H. *Z. Physiol. Chem.* **1939**, *258*, 57.

265. Bruigom, E. S. Thesis, University of Leiden, Feb. 1950.

266. Serdijn, J. Thesis, University of Leiden, Feb. 1985.

267. Skeggs, L. T.; Marsh, W. H.; Kahn, J. R.; Shumway, N. P. *J. Exp. Med.* **1954**, *99*, 275.

268. Page, I. H.; Bumpus, F. M. *Physiol. Rev.* **1961**, *41*, 331.

269. Schwyzer, R. *Pure Appl. Chem.* **1963**, *6*, 265 (see p 281).

270. Havinga, E.; Schattenkerk, C.; Heymens Visser, G.; Kerling, K. E. T. *Recl. Trav. Chim. Pays-Bas* **1964**, *83*, 672, and following papers of this series.

271. De Graaf, J. S.; Jansen, A. C. A.; Kerling, K. E. T.; Schattenkerk, C.; Havinga, E. *Recl. Trav. Chim. Pays-Bas* **1971**, *90*, 301.

272. Bloemhoff, W. Thesis, University of Leiden, Nov. 1974.

273. Laban, J. Thesis, University of Leiden, Oct. 1969.

274. Provó Kluit, P. Thesis, University of Leiden, Oct. 1968.

275. Kolen, A. C. P. M. Thesis, University of Leiden, Nov. 1974.

276. Visser, S.; Roeloffs, J.; Kerling, K. E. T.; Havinga, E. *Recl. Trav. Chim. Pays-Bas* **1968**, *87*, 559.

277. Richards, F. M.; Wyckoff, H. W. In *The Enzymes*, 3rd ed.; Boyer, P. D., Ed.; Academic: New York, 1971; Vol. 4, p 647.

278. Blackburn, P.; Moore, S. In *The Enzymes*, 3rd ed.; Boyer, P. D., Ed.; Academic: New York, 1982; Vol. 15, p 317.

279. Hodges, R. S.; Merrifield, R. B. *J. Biol. Chem.* **1975**, *250*, 1231.

280. Hofmann, K.; Visser, J. P.; Finn, F. M. *J. Am. Chem. Soc.* **1970**, *92*, 2900.

281. Dunn, B. M.; DiBello, C.; Kirk, K. L.; Cohen, L. A.; Chaiken, I. M. *J. Biol. Chem.* **1974,** *249,* 6295.

282. Chaiken, I. M.; Komoriya, A.; Homandberg, G. A. In *Peptides (Proc. 6th Am. Peptide Symp.);* Gross, E.; Meienhofer, J., Eds.; Pierce: Rockford, IL, 1979; p 587.

283. Van Batenburg, O. D. Thesis, University of Leiden, June 1977.

284. Serdijn, J.; Bloemhoff, W.; Kerling, K. E. T.; Havinga, E. *Recl. Trav. Chim. Pays-Bas* **1984,** *103,* 50.

285. Hoes, C. J. T. Thesis, University of Leiden, May 1983.

286. Hoes, C. J. T.; Kerling, K. E. T.; Havinga, E. *Recl. Trav. Chim. Pays-Bas* **1983,** *102,* 140.

287. Tesser, G. J.; Boon, P. J. *Recl. Trav. Chim. Pays-Bas* **1980,** *99,* 289.

288. Merrifield, Bruce. *The Concept and Development of Solid-Phase Peptide Synthesis;* Profiles, Pathways, and Dreams; American Chemical Society, Washington, DC; 1991.

289. Fujii, N.; Yajima, H. *J. Chem. Soc., Perkin Trans. 1,* **1981,** 789–841 (six consecutive papers).

290. Van Kerkwijk, C. P. Thesis, University of Leiden, July 1934.

Index

Index

115

Editing and Indexing: Colleen P. Stamm
Production: Peggy D. Smith
Acquisition: Robin Giroux

Books printed and bound by Maple Press, York, PA

*Paper meets minimum requirements of American National Standard
for Information Sciences—Permanence of Paper for Printed Library
Materials, ANSI Z39.48–1984* ∞